Primary Space Project Research Team

Research Co-ordinating Group

Professor Paul Black (Co-director)
Jonathan Osborne

Centre for Educational Studies

King's College London
University of London
Cornwall House Annexe
Waterloo Road
London SE1 8TX

Tel: 071-872-3094

Professor Wynne Harlen (Co-director)[1]
Terry Russell

Centre for Research in Primary Science
and Technology
Department of Education
University of Liverpool
126 Mount Pleasant
Liverpool L3 5SR

Tel: 051-794 3270

Project Researchers

Pamela Wadsworth (from 1989)

Derek Bell (from 1989)
Ken Longden (from 1989)
Adrian Hughes (1989)
Linda McGuigan (from 1989)
Dorothy Watt ((1986-89)

Associated Researchers

John Meadows
(South Bank Polytechnic)

Bert Sorsby
John Entwistle
(Edge Hill College)

LEA Advisory Teachers

Maureen Smith (1986-89)
(ILEA)

Joan Boaden
Karen Hartley
Kevin Cooney (1986-88)
(Knowsley)

Joyce Knaggs (1986-88)
Heather Scott (from 1989)
Ruth Morton (from 1989)
(Lancashire)

---

[1] Now Director of the Scottish Council for Research in Education, 15 St John St, Edinburgh, EH8 8JR

PRIMARY SPACE PROJECT
RESEARCH REPORT

September 1992

Processes of Life

by
JONATHAN OSBORNE, PAM WADSWORTH
and PAUL BLACK

LIVERPOOL UNIVERSITY PRESS

First published 1992 by
Liverpool University Press
PO Box 147, Liverpool L69 3BX

British Library Cataloguing in Publication Data
Data are available
ISBN 0 85323 178 8

Printed and bound by
Dotesios Limited, Trowbridge, England

CONTENTS

CONTENTS

INTRODUCTION

This introduction provides an overview of the SPACE Project and its programme of research.

The Primary SPACE Project is a classroom-based research project which aims to establish

- *the ideas which primary school children have in particular science concept areas.*

- *the possibility of children modifying their ideas as the result of relevant experiences.*

The research is funded by the Nuffield Foundation and the publishers, Collins Educational, and is being conducted at two centres, the Centre for Research in Primary Science and Technology, Department of Education, University of Liverpool and the Centre for Educational Studies, King's College, London. The joint directors are Doctor Wynne Harlen and Professor Paul Black. The following local education authorities have been involved: Inner London Education Authority, Knowsley and Lancashire.

The Project is based on the view that children develop their ideas through the experiences they have. With this in mind, the Project has two main aims: firstly, to establish (through an elicitation phase) what specific ideas children have developed and what experiences might have led children to hold these views; and secondly, to see whether, within a normal classroom environment, it is possible to encourage a change in the ideas in a direction which will help children develop a more 'scientific' understanding of the topic (the intervention phase).

In the first phase of the Project from 1987 to 1989 eight concept areas were studied:

In the second phase of the Project during 1989 and 1990, a further ten concept areas were studied:

Electricity
Evaporation and condensation
Everyday changes in non-living materials
Forces and their effect on movement
Growth
Light
Living things' sensitivity to their environment
Sound

Earth
Earth in space
Energy
Genetics and evolution
Human influences on the Earth
Processes of life
Seasonal changes
Types and uses of materials
Variety of Life
Weather

Research Reports are usually based on each of these concept areas; occasionally where the areas are closely linked, they have been combined in a single report.

The Project has been run collaboratively between the University research teams, local education authorities and schools, with the participating teachers playing an active role in the development of the Project work.

Over the life-span of the Project a close relationship has been established between the University researchers and teachers, resulting in the development of techniques which advance both classroom practice and research. These methods provide opportunities, within the classroom, for children to express their ideas and to develop their thinking with the guidance of the teacher, and also help researchers towards a better understanding of children's thinking.

The Involvement of the Teachers

Schools and teachers were not selected for the Project on the basis of a particular background or expertise in primary science. In the majority of cases, two teachers per school were involved. This was advantageous in providing mutual support. Where possible, the Authority provided supply cover for the teachers so that they could attend Project sessions for preparation, training and discussion during the school day. Sessions were also held in the teachers' own time, after school.

The Project team aimed to have as much contact as possible with the teachers throughout the work to facilitate the provision of both training and support. The diversity of experience and differences in teaching style which the teachers brought with them to the Project meant that achieving a uniform style of presentation in all classrooms would not have been possible, or even desirable. Teachers were encouraged to incorporate the Project work into their existing classroom organisation so that both they and the children were as much at ease with the work as with any other classroom experience.

The Involvement of Children

The Project involved a cross-section of classes of children throughout the primary age range. A large component of the Project work was classroom-based, and all of the children in the participating classes were involved as far as possible. Small groups of children and individuals were selected for additional activities or interviews to facilitate more detailed discussion of their thinking.

The Structure of the Project

In the first phase of the Project, for each of the concept areas studied, a list of concepts was compiled to be used by researchers as the basis for the development of work in that area. These lists were drawn up from the standpoint of accepted scientific understanding and contained concepts which were considered to be a necessary part of a scientific understanding of each topic. The lists were not necessarily considered to be statements of the understanding which would be desirable in a child at age eleven, at the end of the Primary phase of schooling. The concept lists defined and outlined the area of interest for each of the studies; what ideas children were able to develop was a matter for empirical investigation.

In the second phase of the Project, the delineation of the concept area was informed by the National Curriculum for Science in England and Wales. The concept area was broken into a number of themes from which issues were selected for research. Themes sometimes contained a number of interlocking concepts; in other instances, they reflected only one underlying principle.

Most of the Project research work can be regarded as being organised into two major phases each followed by the collection of structured data about children's ideas. These phases called 'Exploration' and 'Intervention', are described in the following paragraphs and together with the data collection produce the following pattern for the research.

Phase 1a	*Exploration*
Phase 1b	*Pre-Intervention Elicitation*
Phase 2a	*Intervention*
Phase 2b	*Post-Intervention Elicitation*

The Phases of the Research

For the first eight concept areas, the above phases were preceded by an extensive pilot phase. Each phase, particularly the pilot work, was regarded as developmental; techniques and procedures were modified in the light of experience. The modifications involved a refinement of both the exposure materials and the techniques used to elicit ideas. This flexibility allowed the Project team to respond to unexpected situations and to incorporate useful developments into the programme.

Pilot Phase

There were three main aims of the pilot phase. They were, firstly to trial the techniques used to establish children's ideas, secondly, to establish the range of ideas held by primary school children, and thirdly, to familiarise the teachers with the classroom techniques being employed by the Project. This third aim was very important since teachers were being asked to operate in a manner which, to many of them, was very different from their usual style. By allowing teachers a 'practice run', their initial apprehensions were reduced, and the Project rationale became more familiar. In other words, teachers were being given the opportunity to incorporate Project techniques into their teaching, rather than having them imposed upon them.

Once teachers had become used to the SPACE way of working, a pilot phase was no longer essential and it was not always used when tackling the second group of concept areas. Moreover, teachers had become familiar with both research methodology and classroom techniques, having been involved in both of them. The pace of research could thus be quickened. Whereas pilot, exploration and intervention had extended over two or three terms, research in each concept area was now reduced to a single term.

In the Exploration phase children engaged with activities set up in the classroom for them to use, without any direct teaching. The activities were designed to ensure that a range of fairly common experiences (with which children might well be familiar from

---

their everday lives) was uniformly accessible to all children to provide a focus for their thoughts. In this way, the classroom activities were to help children articulate existing ideas rather than to provide them with novel experiences which would need to be interpreted.

Each of the topics studied raised some unique issues of technique and these distinctions led to the Exploration phase receiving differential emphasis. Topics in which the central concepts involved long-term, gradual changes, such as 'Growth', necessitated the incorporation of a lengthy exposure period in the study. A much shorter period of exposure, directly prior to elicitation was used with topics such as 'Light' and 'Electricity' which involve 'instant' changes.

During the Exploration teachers were encouraged to collect their children's ideas using informal classroom techniques. These techniques were:

i   ***Using Log-Books (free writing/drawing)***
Where the concept area involved long-term changes, it was suggested that children should make regular observations of the materials, with the frequency of these depending on the rate of change. The log-books could be pictorial or written, depending on the age of the children involved, and any entries could be supplemented by teacher comment if the children's thoughts needed explaining more fully. The main purposes of these log-books were to focus attention on the activities and to provide an informal record of the children's observations and ideas.

ii   ***Structured Writing/Annotated Drawing***
Writing or drawings produced in response to a particular question were extremely informative. Drawings and diagrams were particularly revealing when children added their own words to them. The annotation helped to clarify the ideas that a drawing represented.

Teachers also asked children to clarify their diagrams and themselves added explanatory notes and comments where necessary, after seeking clarification from children.Teachers were encouraged to note down any comments which emerged during dialogue, rather than ask children to write them down themselves. It was felt that this technique would remove a pressure from children which might otherwise have inhibited the expression of their thoughts.

iii   ***Completing a Picture***
Children were asked to add the relevant points to a picture. This technique ensured that children answered the questions posed by the Project team and reduced the possible effects of competence in drawing skills on ease of expression of ideas. The structured drawings provided valuable opportunities for teachers to talk to individual children and to build up a picture of each child's understanding.

iv   ***Individual Discussion***
It was suggested that teachers use an open-ended questioning style with their children. The value of listening to what children said, and of respecting their responses, was emphasised as was the importance of clarifying the meaning of

words children used. This style of questioning caused some teachers to be concerned that, by accepting any response whether right or wrong, they might implicitly be reinforcing incorrect ideas. The notion of ideas being acceptable and yet provisional until tested was at the heart of the Project. Where this philosophy was a novelty, some conflict was understandable.

In the Elicitation which followed Exploration, the Project team collected structured data through individual interviews and work with small groups. The individual interviews were held with a random, stratified sample of children to establish the frequencies of ideas held. The same sample of children was interviewed pre- and post-Intervention so that any shifts in ideas could be identified.

Intervention Phase

The Elicitation phase produced a wealth of different ideas from children, and produced some tentative insights into experiences which could have led to the genesis of some of these ideas. During the Intervention, teachers used this information as a starting point for classroom activities, or interventions, which were intended to lead to children extending their ideas. In schools where a significant level of teacher involvement was possible, teachers were provided with a general framework to guide their structuring of classroom activities appropriate to their class. Where opportunities for exposing teachers to Project techniques had been more limited, teachers were given a package of activities which had been developed by the Project team.

Both the framework and the Intervention activities were developed as a result of preliminary analysis of the Pre-Intervention Elicitation data. The Intervention strategies were:

(a) **Encouraging children to test their ideas.**
It was felt that, if pupils were provided with the opportunity to test their ideas in a scientific way, they might find some of their ideas to be unsatisfying. This might encourage the children to develop their thinking in a way compatible with greater scientific competence.

(b) **Encouraging children to develop more specific definitions for particular key words.**
Teachers asked children to make collections of objects which exemplified particular words, thus enabling children to define words in a relevant context, through using them.

(c) **Encouraging children to generalise from one specific context to others through discussion.**
Many ideas which children held appeared to be context-specific. Teachers provided children with opportunities to share ideas and experiences so that they might be enabled to broaden the range of contexts in which their ideas applied.

(d) **Finding ways to make imperceptible changes perceptible.**
Long-term, gradual changes in objects which could not readily be perceived were problematic for many children. Teachers endeavoured to find appropriate ways of making these changes perceptible. For example, the fact that a liquid

could 'disappear' visually yet still be sensed by the sense of smell - as in the case of perfume - might make the concept of evaporation more accessible to children.

(e) **Testing the 'right' idea alongside the children's own ideas.**
Children were given activities which involved solving a problem. To complete the activity, a scientific idea had to be applied correctly, thus challenging the child's notion. This confrontation might help children to develop a more scientific idea.

(f) **Using secondary sources.**
In many cases, ideas were not testable by direct practical investigation. It was, however, possible for children's ideas to be turned into enquiries which could be directed at books or other secondary sources of information.

(g) **Discussion with others.**
The exchange of ideas with others could encourage individuals to reconsider their own ideas. Teachers were encouraged to provide contexts in which children could share and compare their ideas.

In the Post-Intervention Elicitation phase the Project team collected a complementary set of data to that from the Pre-Intervention Elicitation by re-interviewing the same sample of children. The data were analysed to identify changes in ideas across the sample as a whole and also in individual children.

These phases of Project work form a coherent package which provides opportunities for children to explore and develop their scientific understanding as a part of classroom activity, and enables researchers to come nearer to establishing what conceptual development it is possible to encourage within the classroom and the most effective strategies for its encouragement.

The Implications of the Research

The SPACE Project has developed a programme which has raised many issues in addition to those of identifying and changing children's ideas in a classroom context. The question of teacher and pupil involvement in such work has become an important part of the Project, and the acknowledgement of the complex interactions inherent in the classroom has led to findings which report changes in teacher and pupil attitudes as well as in ideas. Consequently, the central core of activity, with its data collection to establish changes in ideas should be viewed as just one of the several kinds of change upon which the efficacy of the Project must be judged.

The following pages provide a detailed account of the development of the Processes of Life topic, the Project findings and the implications which they raise for science education.

The research reported in this and the companion research reports, as well as being of intrinsic interest, informed the writing and development with teachers of the Primary SPACE Project curriculum materials, to be published by Collins Educational.

1:Previous Research-A Review

Processes of life such as growth, reproduction, movement, feeding, excretion, respiration and sensitivity are fundamental to any biological knowledge and the core concept of *living thing*. Studies of the development of the child's understanding of the processes of life have invariably focused on two aspects. Initial research examined animistic thinking by children and their concept of life and the criteria they deploy for establishing whether an object is 'living' or 'not alive'. Later studies examined the child's perception of the inside of the body and the processes of life themselves. Somewhat surprisingly most of this work has been undertaken by those working in the field of psychology, nutrition and nursing and does not appear to be generally well known amongst science educationalists. Good summaries can be found in Carey (1985) and Mintzes (1984).

Living and Non-living

Perhaps the most well known of these studies are those undertaken by Piaget (1929) who established a framework based around the criterion of movement. Piaget's technique was to use the clinical interview and present the subject with an object and ask the question, 'Is it alive?' and if the child's answer was 'Yes', he asked 'How do you know?'. From his results, he distinguished the following stages of development in the concept of what constituted a 'living object'.

Stage 0	*No Concept*	Random judgements or inconsistent or irrelevant justifications
Stage 1	*Activity*	Things that are active in any way (including movement) are alive
Stage 2	*Movement*	Only things that move are alive
Stage 3	*Autonomous Movement*	Things that move by themselves are alive
Stage 4	*Adult Concept*	Only animals (or animals and plants) are alive

Piaget's early work was developed into a standardised interview procedure by Russell and Denis (1939) and the area has been the focus of many replication studies, the most notable being that of Laurendeau & Pinard (1962). They tested 500 subjects between the ages of 4 and 12 and agreed with Piaget's conclusions apart from finding no evidence for a distinction between stages 1 & 2.

Further studies in the field have generally given results which support this interpretation (see Jahoda (1958), Looft and Bartz (1969) for reviews). That this interpretation is open to question has come from studies which have adopted a different methodology and attempted to focus on what children conceive the 'attributes of life' to be and how these develop. Such studies have opened a rich field for exploration of which the research here merely represents a continuation.

One of the earliest studies was undertaken by Looft (1974) who asked children "Does a frog breathe or need air?" "Does a chair need food or nutrition?" "Do automobiles reproduce or make more things just like themselves?" Looft also asked his subjects if the items used in the question were 'alive'. His important discovery was that although some students could correctly assign all of these objects to 'living' or 'not living', there was a lack of a full understanding of the attributes of life. Such work does not contradict the earlier studies and could be considered to supportive in that it shows that children are clearly not using the 'attributes of life' as the prime criterion for deciding the issue of whether an object is alive/not alive. However, it does reveal a disparity between a child's and adult's concept of an animate object, and that children lack domain-specific biological knowledge.

A further study, by Smeets, investigated whether children were capable of correctly attributing six life traits (die, grow, feel, hear, know, talk) to animate and inanimate objects. He found that these processes of life were often incorrectly attributed to inanimate objects.

Working in a different tradition, in which the conceptual development of children is studied from a psychological perspective, Carey (1985) chose to examine the development of children's understanding of alive/not alive and their accompanying biological knowledge between the ages of 4 and 12. Carey argues that the use of the framework 'alive', 'not alive' is simplistic forcing a categorisation which is not meaningful to the child. The inevitable failure to categorise is due to a lack of biological knowledge. She argued that such knowledge gradually improves between these ages resulting in a domain-specific restructuring, and it is this restructuring which results in the improvement of children's abilities to respond to the question of whether an object is alive/not alive. In a similar study to Smeets, she specifically chose unfamiliar animals. e.g aardvarks, dodos, garlic presses, clouds. She tested 9 subjects each from ages 4,5,7 and adults and found that at no age were animal properties attributed to inanimate objects. Hence her results contradict the findings of Smeets.

Her most striking finding was the under-attribution of animal properties to animals other than people, in particular breathing and eating, which led to a failure to attribute these properties to all animals. Carey postulated three mechanisms for children's reasoning:- deductive inference based on some narrow concept of an animal; the application of a definition which would involve checking for the component parts associated with the process i.e mouth for eating, nose for breathing, and inductive projection based on comparison with humans. She concluded that, although all three types of reasoning contribute, the primary basis of their reasoning was the third mode- inductive projection. Her argument was that the evidence showed that there was a major restructuring of domain specific knowledge by the child reached the age of 10. This enabled the child to conceptualise the human body in terms of an integrated functioning of internal organs and perceive other living things in similar terms.

Lucas et al (1979) identify a number of methodological errors in these studies. They argue that the increasing facility with age may just reflect an increasing familiarity with the everyday objects used. Secondly there are conceptual difficulties with the 'attributes of life' used which are strongly biased towards humans and ignore plants. The consequence is a tendency to over-rely on an anthropomorphic framework which would result in category errors. Finally, like Carey they argue that the method of interviews used force criteria on the children which are not necessarily those which the child would spontaneously use.

Lucas et al's response was to use a technique which avoided some of these mistakes by showing children a black and white photograph of an 'object' which had been found on a beach. Children were then asked 'How could you find out if the object was alive? Write down as many ways as you can think of' The study was done with 944 students from Grade 2 (age 6) to Grade 10 (Age 14). Their research identified five broad categories which students spontaneously used for establishing whether the object was living - expert advice, external structure, internal structure, physiological functions and behaviour. No children used one category only and although the work confirmed the use of the criterion of spontaneous movement found in earlier studies, the most revealing aspect was the lack of predominance of this criterion. At all grade levels, more than 40% of pupils suggested a criterion based on external structure. In addition, an increasing proportion at higher grade levels used a criterion based on internal structure and/or physiological functions. The authors argue that previous work has ignored the 'richness of children's responses to a highly complex question' and that the context of the data gathering can have an important effect on the nature of the response obtained.

In summary, early studies would seem to have attempted to reduce the child's view of the world to a description which later work has shown to be simplistic. The evidence is that there are several facets to the criteria deployed by children, not least of which is their biological knowledge.

Human internal organs

The most well-known study is that of Gellert (1962) in which she asked 96 children, age 4 to 16, to list what they have inside them. In an extensive study, she investigated where children thought the major organs are found inside the body, what the role of each is and what would happen if if one lacked such an organ. The overriding conclusion of her study was the development in knowledge between infants/lower juniors (5-8) and upper juniors (9+). The former group came up with approximately 3 things inside people whilst the latter were able to list 8. The younger group predominantly think in terms of what they have seen put in, and coming out i.e. food and blood whilst the older group add a wide variety of internal organs. Another important finding was that when asked, "What do you think is the most important part of the body?", the younger group responded with external parts e.g. hair, nose, feet, eyes whilst by age 10, children respond with internal bodily organs.

Gellert also showed that young children's understanding of defecation is one which sees the process of social necessity, necessary so that we will not get too full or burst. Only when children reached the age of 13/14 did they see the process as the elimination of waste or noxious substances by the body.

Further studies undertaken since then have confirmed this analysis (Wellman & Johnson (1982), Contento (1981). In particular what they show is a lack of understanding by very young children, age 5-6, of what happens to food. Most know that it goes to the stomach but imagine that it stays there unchanged or is broken into smaller bits. Contento's work showed a strong relationship between Piagetian stages and such understanding. All the children at a pre-operational level considered food to remain unchanged when eaten, whereas children at a concrete level recognised that food changes but the majority did not know how.

Gellert's study clearly showed that the heart was the first internal organ that children were aware of, partly because it has a clearly detectable presence in that it 'beats'. By the age of 10 or 11 well over half this age group realised that the heart is a pump and circulates blood around the body. Again very few of the younger children under 7 in Gellert's study had heard of lungs or could begin to explain their function. Only by the age of 10 did they show an understanding of the role of the lungs in exchanging gases and the circulation of air/oxygen to the rest of the body.

Crider (1981) has attempted to place some kind of theoretical framework on these descriptive lists which one author has described as the 'conceptual ecology' of the classroom (Driver, 1989). She argues that when the young child comes to know an internal organ, each is assigned a single function e.g the lungs are for breathing. From such ideas the child moves to perceiving an inter-relatedness of the organs which are perceived as containers with channels connecting them. The final stage involves the development of a particulate understanding which sees matter such as food as being reducible to a microscopic level at which it can be transported around the body. Crider argues that this is achieved by the age of 11 for many pupils but in view of the research on children's understanding of the particulate nature of matter (Brook, Briggs and Driver, 1984) which shows that the majority of the children are incapable of understanding such an idea, this argument must be open to question.

Johnson & Wellman (1982) also conducted a study of children's understanding of the nature and location of the brain. Their study looked at what children perceived to be its function and what activities require a brain. In summary, awareness of the brain as an internal organ begins at age 4 where its function is recognised for thinking. What was not recognised was that involuntary motor acts such as walking, coughing, sleeping required activity by the brain. Children of age 5 saw the brain as being autonomous from a whole range of body parts e.g eye, mouth, ear, but by age 10 nearly 80% saw the brain as helping the body parts. Essentially young children see the brain as a mental organ which has no specific physiological function. Children's understanding of nerves consequently is very limited other than that they are an integral part of the body with no specific function. Only after age 9 were some children able to assign them a specific function related to conducting messages, controlling activity or sensing pain.

One notable point that emerges from Johnson & Wellman's study is the effect of instruction about the brain to a group of 11 year old children. Their research took place before this group studied a unit on the brain. They investigated their understanding after the unit had been taught and found that the teaching sequence had had absolutely no effect on their learning.

Other Processes

The other two processes extensively studied are birth and death. The two most significant studies of birth are by Bernstein and Cowan (1975) and Goldman and Goldman (1982). Bernstein and Cowan classified children's progression into 6 levels of understanding from that of the youngest children, level 1, whose explanation for babies was that babies had always existed, to children at level 6, who explained conception in terms of the fertilisation of the egg and the combination of genetic material. Level 4, at which the child recognises that the 'seed' from the father is united with the egg from the mother, is the one that is independent of animism and artificialism. Goldman and Goldman's cross-cultural study of North American, English, Australian and Swedish children revealed that English children were significantly weaker at attaining a level 4 understanding by age 11. It is of course interesting to note that Swedish children were the best and found to be four years ahead of their peers in other countries.

North America	England	Australia	Sweden
80	63	87	97

Table 1.2: Percentage of children attaining a level 4 understanding
of the process of reproduction by age 11. (Goldman & Goldman, 1982)

Carey argues that the data show clearly that young children see the production of babies only in terms of the intentionality of their parents and have no knowledge of the function of the body in the process. By age 10 they make a clear distinction between the role of the body and the role of the parents.

The problems posed by death in families and the effect on children have led to some very extensive research by psychologists. Again Carey (1985) summarises much of the wide-ranging literature. Psychologists essentially identify three phases. In the first stage (age 5 and under), children have no concept of the cessation of biological function and death is seen in terms of a separation which is neither final or inevitable. In the second stage, the child now recognises the finality of death but sees death as being caused by an external agent e.g. guns, knives, 'Father Death', poisons. In the final stage, which occurs for most children around age 9 or 10, death is seen as an inevitable biological process . Whilst death cannot be separated from the human and emotional perspective, Carey argues that it is the irreversibility of the process which leads to the emotional impact and that children's level of understanding of death by age 9/10 shows that they have developed the biological knowledge to appreciate the significance of death from an adult perspective as such an irreversible process.

Conclusions

Clearly the existing body of research in this domain is extensive but, as noted, not well known to science educators and much of it pre-dates the work of the 'alternative conceptions' movement. Many of the studies have attempted to place their findings within the context of a Piagetian developmental perspective i.e. pre-operational, concrete and formal. Carey (1985) argues that there is little to be gained by such a process because such a structure is a description of children's logic which fails to accurately interpret the nature of children's thinking and secondly it 'commits one to the claim that there is something which limits the understanding of digestion or the origin of babies.' Instead she develops a case that the evidence suggests a restructuring of domain-specific knowledge which enables a shift in conceptualisation of the processes of life.

Whilst it it is not the intention to enter into this debate here, the research reported in this document is an attempt to explore the demands of the English & Welsh national curriculum and add to this body of knowledge in a form which is hopefully more accessible to the large number of primary teachers who will be confronting the teaching of these scientific concepts. The research reported adopted a constructivist perspective. Hence it used many of the techniques used in previous research and adapted others to elicit children's ideas and so yield a broad picture of children's intuitive understandings of these biological concepts. This elicitation was followed by an intervention process which provided an opportunity to generate conceptual conflict with children's existing ideas. Finally, a second elicitation was undertaken to examine what changes had occurred in children's understanding of the processes of life.

References

Bernstein, A.C & Cowan, P.A. (1975) Children's Concept of How People Get Babies. *Child Development,* 46, 77-91.

Brook, A., Briggs, H. & Driver, R. (1984.) *Aspects of Children's Understanding of the particulate nature of matter.* Children's Learning in Science Project. Centre for Studies in Mathematics and Science Education. The University of Leeds.

Carey, S. (1985.) *Conceptual Development in Childhood.* Cambridge: MIT Press.

Driver, R. (1989) Student's conceptions and the learning of science. *International Journal of Science Education,* 11, 481-490.

Gellert, E. (1962) Children's conception of the content and functions of the human body. *Genetic Psychology Monographs,* 65, 291-411.

Goldman, R.J. and Goldman, J.D.G. (1982) How children perceive the origin of babies and the roles of mothers and fathers in procreation: A cross-national study. *Child Development,* 53,491-504.

Jahoda, G. (1958) Child animism: 1. A critical survey of cross-cultural research. *J.Soc.Psychol.,* 47, 197-212.

Laurendeau, M. and Pinard, A. (1962) *Causal Thinking in the Child: A Genetic and Experimental Approach.* New York: International Universities Press.

Looft, W.R. (1974) Animistic thought in children: Understanding the "living" across its associated attributes. *J. Genet. Psychol.,* 124, 235-240.

Looft, W.R., & Bartz, W.H., (1969) Animism Revived. *Psychol. Bull.,* 71, 1-19.

Lucas, A.M., Linke, R.D. & Sedgwick, P.P. (1979) School children's criteria for "Alive": A content analysis approach.*The Journal of Psychology,* 103, 103-112.

Mintzes, J.J. (1984) Naive Theories in Biology: Children's Concepts of the Human Body. *School Science and Mathematics,* 84, 7, pp 548-556.

Piaget, J. (1929) *The Child's Conception of the World.*New York. Harcourt Press.

Russell, R.W. & Dennis, W. (1939) Studies in animism. A standardised procedure for the investigation of animism. *J. Genet. Psychol.,* 55, 389-400.

Smeets, P.M. (1973) The animism controversy revisited: A probability analysis. *J.Genet. Psychol.,* 123, 219-225.

2:Methodology

Sample

a. Schools

Six schools from the London area were chosen for this research from three local authorities (Inner London, Newham and Barnet). One teacher from each school participated in the project. Each school was allocated to one member of the research team[1] who worked closely with the teacher throughout the research phase.

The majority of the schools were selected by the research officer who had already been working in the locality providing support to primary schools in the development of primary science work in her previous post.

b. Teachers

Most of the teachers invited to participate in the project were those known to the researchers from the previous work. This was advantageous in providing a pre-existing relationship and link between researcher and teachers which could be developed. Teachers were able to use this relationship to express their uncertainties about the work and ask for clarification. Unfortunately, the local authority was unable to release any of the teachers due to the difficulties experienced during this phase in obtaining any supply cover in the London area. This meant that all meetings had to take place during the teachers' own time after school, and this had the effect of curtailing the extent of the teacher contribution to the research on this topic.

The teacher's normal style of working varied, between individuals who made sole use of classrooms organised around groups using a topic approach and an 'integrated' day, and those who preferred to keep the class working together on a common theme. Teachers were encouraged to integrate the activities into their existing mode of working as there was a limitation to the amount of change of teaching style that could be expected of them.

---

[1]   The research reported here was undertaken by the authors, Pam Wadsworth (full-time) and Jonathan Osborne (part-time) during 1989.

Many of the difficulties experienced and expressed by teachers with a topic are associated with a lack of confidence in their own understanding of the background science. In particular, this results in an concern about the level of understanding that it would be reasonable to expect a child to achieve. Whilst teachers understood that the research project was attempting to provide some insight into the latter question, it was clear that the degree of uncertainty was a source of anxiety for teachers.

Names of the participating schools are provided in Appendix 1.

c. Children

Despite the limitation to a particular locale, the schools used reflect the wide variation seen in the London area between schools based in deprived areas and those with a substantial middle-class catchment area. Hence the children used in the sample represent children with a wide range of ability and ethnic background. All children in the classes of the participating teachers were used for the pre- and post-intervention elicitation activities. Inevitably there were some children who were not present for both phases of the activity and the data collected from these children have not been used.

For the purpose of analysis, the children have been grouped by age into infants (5-7), lower juniors (8-9) and upper juniors (10-11). In case of any doubt surrounding the particular grouping of a child, the year of schooling was used to decide the appropriate cohort for a child. Data was generally obtained by individual interview though some of the data from lower and upper junior children was obtained through written responses.

d. Liaison

During the data-collection phase of the project, the research was conducted by two people working part-time with the schools and the relevant teachers. Each member of the team was allocated a particular school. The researchers would meet on a regular basis to plan and co-ordinate the research, exchange information and develop materials.

The Research Programme

Classroom work on the topic of 'processes of life' took place over a relatively long period in the school year which can be summarised as follows.

Pilot Exploration	Sept 89
Pre-Intervention Data Collection	Oct 89
Intervention	Nov 89
Post-Intervention Data Collection	Dec-Jan 90

The pilot exploration phase was based on interviews with a small number of children (20). These interviews used a wide range of questions to explore the nature of children's understanding of the processes of life and associated concepts. In addition, drawings and answers to written questions were employed to examine how valuable and reliable such sources were for eliciting children's meanings and understanding. The exploratory nature of this phase was required to supplement what little literature there was available on the nature of young (5-11) children's understanding of this topic and to explore how suitable the questions were for eliciting children's understanding of the concepts. Some of the questions devised for probing children's ideas were modifications of methods that had been used previously by other researchers. At the end of this phase, the data were examined to determine which were the most valuable lines of approach for eliciting children's ideas about this topic. The other valuable feature of this phase was that it provided time for developing a relationship with the teacher and the children so that they could become accustomed to the mode of working required.

Essentially, the classroom elicitation techniques were refined by the pilot process and the experience provided an opportunity for teachers and researchers to develop familiarity with the material and with each other. Data on children's ideas were then collected from children in classrooms using the selected activities. These questions and activities are shown in Appendix 2. The main methods of elicitation relied on a mixture of interviews, written answers and children's drawings. All the data from infant children were collected by interview and drawings as these children found it very difficult to provide written answers to questions.

The intervention activities were designed in consultation with the teachers and from an examination of the data collected previously. The data suggested several areas of interest for possible conceptual development and a framework of activities was designed which could be used by children to test their own ideas and explore their thinking in this domain. This was not presented as a prescriptive framework, but simply as a range of exercises and activities which could be used by children. Teachers and children were free to try other lines of investigation they wished to pursue. After the completion of the intervention phase, another set of elicitations was used with the children based on the same questions to those used in the elicitation prior to the intervention.

Defining 'The Processes of Life'

Any attempt to develop a child's concepts needs to be based on a definition of what a preferred understanding would be. In the earlier research, a list of concepts was compiled by the team to provide a map of ideas considered an *a priori* necessity for the development of the scientist's world view. However, in this instance, the National Curriculum Order had been published and the framework of the research changed. The Order defined, in a set of attainment targets, learning objectives for children to achieve through the age range in a progressive, developmental fashion. Whilst the Order and their articulation of the targets within it are open to debate, they represented at the time, the standard objectives that many teachers would be using for their teaching. Hence the decision was made to adopt these statements as guidelines of what it might be reasonable for a child to be expected to know. This does not imply that the team necessarily accepted these statements as reasonable expectations but they did constitute a set of aims for many teachers and their children. Therefore the research set out to ask whether they were reasonable expectations.

The National Curriculum then was defined in terms of a set of attainment targets and programmes of study. The attainment targets (Fig 2.1) represented assessment objectives on a 10 point scale. An able infant is expected to achieve level 3 by age 7 whilst an average child would achieve level 2. A able junior should achieve level 5 by the age 11 whilst an average child level 4. The programmes of study (Fig 2.2) merely defined the set of experiences that should enable the attainment targets to be achieved.

The purpose of this list is to provide a framework or point of reference for the research where these statements represent a collection of ideas that children *may* develop by age 11. The principal difference between this research and earlier work on light and electricity, is that this is an externally defined list. One of the aims of the research is to examine to what extent, as a consequence of the experiences that were provided by this research programme, such ideas develop in children and at what ages.

Level	Attainment Target
1	Pupils should: • be able to name or label the external parts of the human body/plants, for example, *arm, leg/flower, stem*
2	• know that living things reproduce their own kind • know that personal hygiene, food, exercise, rest and safety, and the proper use of safe medicines are important. • be able to give a simple account of the pattern of their own day.
3	• know that the basic life processes: feeding, breathing, movement, behaviour, are common to human beings and the other living things they have studied. • be able to describe the main stages of the human life cycle.
4	• be able to name the major organs and organ systems in flowering plants and mammals. • know about the factors which contribute to good health and body maintenance, including the defence systems of the body, balanced diet, oral hygiene and avoidance of harmful substances such as tobacco, alcohol and other drugs. • understand the process of reproduction in mammals • be able to describe the main stages of flowering plant reproduction.
5	• know that living things are made up from different kinds of cells which carry out different jobs. • understand malnutrition and the relationships between diet, exercise, health, fitness and circulatory disorders. • know that in digestion food is made soluble so that it can enter the blood. • understand the way in which microbes and lifestyle affect health • be able to describe the functions of the major organ systems.

Fig 2.1: Attainment Target 1-5 of the English & Welsh National Curriculum (DES, 1989)[1]

The programmes of study were as follows.

---

[1] Since the publication of this Order, a revised publication has been produced by the Department for Education in 1991. The work reported here was based on the original Order. The summary and conclusions of this work are based on the new order (DES, 1991)

Key Stage 1[1]	Children should be finding out about themselves, developing their ideas about how they grow, feed, move and use their senses and about the stages of human development. Using suitable books, pictures and charts, they should be introduced to ideas about how they keep healthy through exercise and personal safety. Children should be introduce to the role of drugs as medicines.
Key Stage 2	Children should investigate some aspects of feeding, support, movement and behaviour in relation to themselves and other animals. They should be introduced to the functions of the major organs systems and to basic ideas about the processes of breathing, circulation, growth and reproduction. They should explore ways in which good health can be promoted in relation to their own daily routine, using a range of secondary sources chosen by the teacher. They should be introduced to the fact that while all medicines are drugs, not all drugs are medicines; and they should begin to be aware of the catastrophic effect on health resulting from an abuse of drugs. They should investigate the effects of physical factors on the rate of plant growth, for example, *light intensity, temperature and the amount of fertiliser.*[2]

Fig 2.2: Programmes of Study for the English & Welsh National Curriculum in Science at Key Stage 1 & 2.

These ideas also provide a framework for examining children's ideas allowing three questions to be addressed.

a)   How different were the conceptions held by many children from such a framework and how disparate were their ideas?

b)   What development was observable in children's ideas across the age range?

c)   What potential did the planned intervention have for the development of children's ideas towards the scientist's view?

---

[1]   The term key stage refers to the period of education. Key stage 1 is from age 5-7 (two years) and Key stage 2 is from age 7-11 (four years).

[2]   Italicised parts of these documents are provided only as exemplars.

This list was also used as a reference point for the development of the intervention. Given such a framework of objectives, the intervention task was to develop activities which would assist the formation of a fuller understanding of this domain in children. The activities were devised using simple materials familiar to children. Their primary role was to provide a focus for discussion of children's thinking and to challenge their existing ideas. Other considerations in designing activities were that the materials should be simple, easy to manipulate and safe to handle.

3: Children's Ideas about the Processes of Life

This chapter presents a qualitative picture of young children's thinking about the processes of life as found during the elicitation phase with individual or small groups of children. The elicitation of their thinking was carried out by teachers and researchers using a subset of the activities employed in the pilot phase. The picture presented is based on a sample of 75 children (29 infants, 23 Lower Juniors and 23 Upper Juniors). As such it does not claim to be comprehensive but simply presents a broad sketch of the typical thinking of many children in this domain. Details of the elicitation activities are provided in Appendix 2.

The elicitation consisted of a mix of activities. In some cases children were asked to provide written responses and drawings and in other cases they were interviewed. For instance, all the infant children were interviewed for all the questions because of the difficulty that many had in writing. However, with the upper and lower juniors, written responses were used wherever possible to minimise the amount of interviewing required. The elicitation of children's ideas about the living/non-living nature of a range of objects was done by interview with all children because it was thought necessary to explore their ideas on this issue as fully as possible.

The elicitation was designed to explore five areas of children's knowledge and understanding. These can be typified as

a.   What choices and actions are required for healthy living?
      (Appendix 2, question 1/2/8/9/10)
b.   What knowledge of the human body did children have?
      (Appendix 2, question 5/6/7/12/13)
c.   What processes are performed by components of the body?
      (Appendix 2, question 4/11/13)
d.   What understanding did children have of the concept of 'living thing'?
      (Appendix 2, question 15)
e.   What was the child's knowledge of plants and their parts?
      (Appendix 2, question 14)

Hence the data are described under these general headings. A full analysis of the data can be found in Chapter 5.

What choices and actions are required for healthy living?

Health education is a topic undertaken to varying degrees in schools and a focus of much attention in the media. Articles and features in popular magazines, newspapers and the media often address issues of concern in this domain e.g. smoking, dietary fibre, exercise. This concern has been reflected in the curriculum with a greater emphasis placed on the development of attitudes and knowledge in schoolchildren conducive to healthy living. One simple question that arises is whether it is possible for children to understand the causal relationship underpinning their choice of actions if they do not possess a basic knowledge of their organs and bodily systems. For instance, the importance of dietary fibre has little significance if a child has no knowledge of intestines. Similarly, the effect of smoking on the alveoli of the lungs leading to emphysema and other diseases is unlikely to be understood. Despite these limitations, there is undoubtedly an element of conditioning generated by constant exposure to arguments for actions and choices for living. The question then for this research was - to what extent had such arguments been assimilated by children, and to a lesser degree - to what extent were they understood?

The first question to explore this understanding was a question which presented children with a range of foods and asked them to ring those which they considered to be healthy. The foods were categorised by the researchers into three groupings - healthy[1], indeterminate[2] and unhealthy[3] and the pupils' responses analysed. The results are displayed in Fig 3.1.

What the data show is that more than two thirds of the healthy foods were recognised by the children and that there is little variation between the age groups. Infants were slightly better at discriminating healthy foods, correctly identifying a larger percentage of those presented, but were inferior at discriminating unhealthy foods. One possible explanation is that there is a tendency on the part of infant children to indicate that *any* food is healthy and this may not reflect a greater knowledge or familiarity. The percentages for the indeterminate group of foods (bread, meat and potatoes) selected by infants as 'healthy' were also quite high. This may reflect a view that such foods are essential foods and hence healthy and a lack of knowledge of the complications attached to these e.g the eating of too much fatty meat, the lack of dietary fibre in white bread and the excessive fat in potatoes in the form of chips.

---

[1]   Healthy foods were considered to be lettuce, orange, apples, juice, rice

[2]   Indeterminate foods were meat, bread and potatoes. Whilst they can form part of a healthy diet, fatty meat, white bread and chips all have particular health problems associated with their consumption.

[3]   Unhealthy foods were sugar, chips, coke, burgers, crisps, sweets and biscuits.

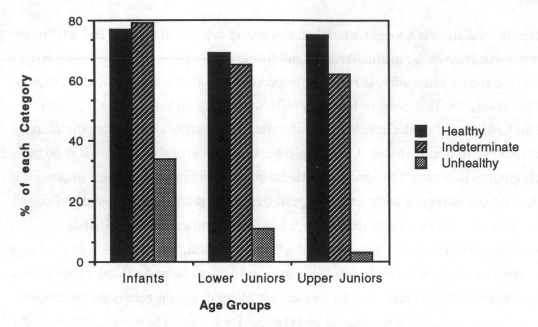

Fig 3.1: Chart showing the percentage of all the possible foods in each group identified as 'healthy' by each age group.

What this data does clearly show is that these children had little difficulty in making a judgement of what constituted a healthy food. This view was supported by another question which asked children to draw on two empty plates - a healthy meal and an unhealthy meal. Typical responses are shown in Fig 3.2 and Fig 3.3.

These responses were typical in that children of all ages were capable of making a satisfactory distinction. Older children tended to produce better drawings, mention more foods and discriminate between healthy and unhealthy foods more effectively. For upper juniors, the foods most commonly drawn as healthy in order of popularity were carrots, vegetables, lettuce and fish. Similarly for lower juniors they were vegetables, carrots, peas and fish.

However, the order for infants was chips, bread, fish and fruit. As such, this shows a marked distinction between the infants and the other two groups in the failure to recognise that chips are generally conceived of as being a relatively unhealthy way of consuming potatoes. The numbers of infants who gave these responses were also much lower than with the lower or upper juniors. Both of these facts would suggest that the notion of what constitutes a healthy food is not so clearly established amongst this group. This is also the first of many instances where a clear difference emerged between infants and the other two groups of children.

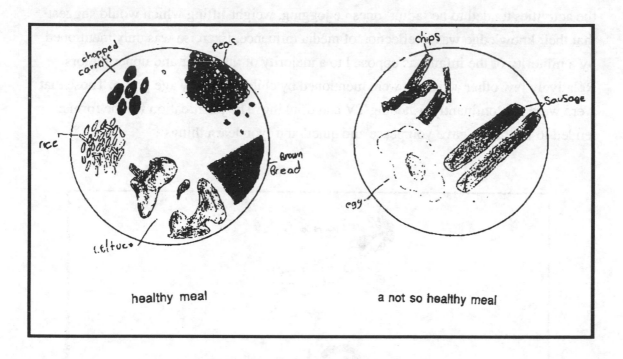

Fig 3.2: Leah-age 9

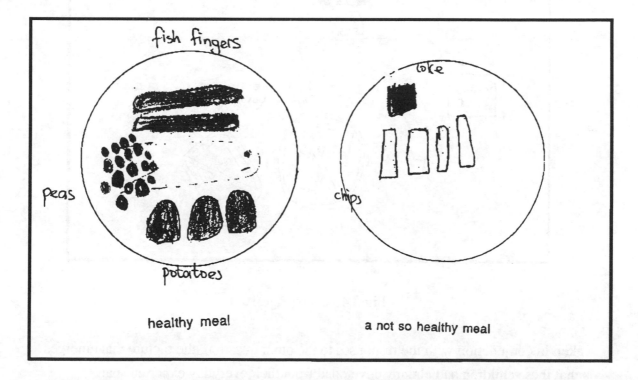

Fig 3.3: Muhammed-age 6

Children were also asked to draw four things that are to do with keeping healthy.
Typical responses are shown in Fig 3.4 and Fig 3.5. Overwhelmingly, pupils of all
ages drew food in response to this question. The second most popular choice was some
indication of exercise or sleeping as being a healthy activity. It was notable though that

the activities tended to be 'adult' ones i.e jogging, weight-lifting which would suggest that their knowledge was a reflection of media influence. Exercise was only mentioned by a minority of the infants as opposed to a majority of the lower and upper juniors. Relatively few other activities were mentioned by children of any age group. Those that were were predominantly watching TV and drinking. The justification for the former tended to be that 'it gave you peace and quiet' and 'you learn things'.

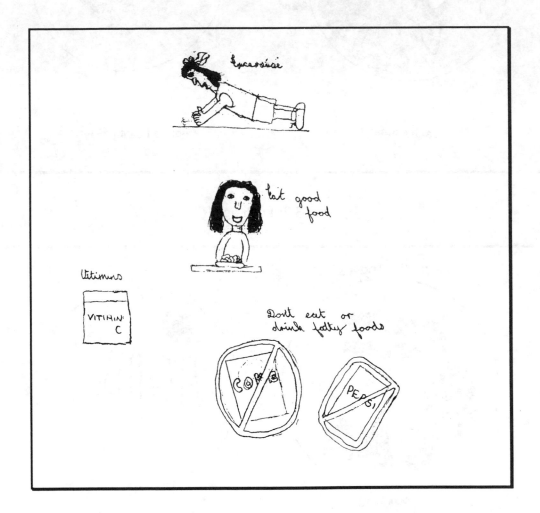

Fig 3.4 Alexis: Age 10

Taken in conjunction with the responses to the other question, the picture that emerges is that these children had already developed from their everyday experience an association of healthy living with certain types of food and were able to identify what constitutes such a food. Consequently it was unlikely that any intervention would have done anything to improve children's ability to discriminate in this domain until it provided children with a more refined knowledge of the variety of types of food and their associated functions.

The final question in this area of children's understanding of health explored whether children had an understanding of the more psychological aspect of keeping healthy associated with feelings and actions. Children were shown a list of examples (Running, feeling happy, swimming, playing with friends, sleeping, eating, laughing, reading, arguing, watching television, smoking and fighting) and asked which of these they associated with keeping healthy.

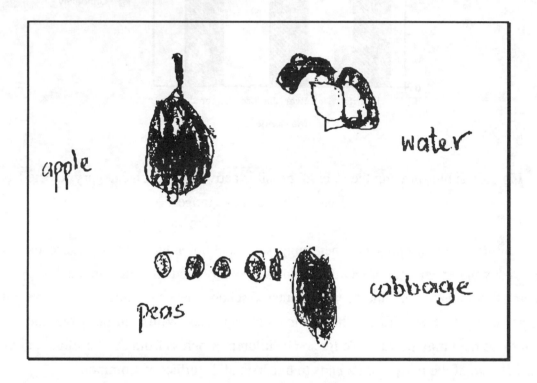

Fig 3.5 Sarah: Age 6

The activities were grouped into three groups by the researchers. These were 'healthy' (running, feeling happy, swimming, playing with friends), 'indeterminate' (reading, laughing, arguing and eating) and 'unhealthy' (smoking, watching television and fighting). Fig 3.6 summarises the data showing the extent to which children were capable of making the same choices as the researchers.

The data would again suggest that the infants were better than lower or upper juniors at discriminating 'healthy' activities. However, the fact that they selected a significant percentage of each of the 'unhealthy' activities as being 'healthy' would imply that there is a tendency by infants to select *any* activity as being 'healthy'. This would explain why they apparently seem to be more successful at discriminating healthy activities and why they state that a large number of the indeterminate activities are 'healthy'.

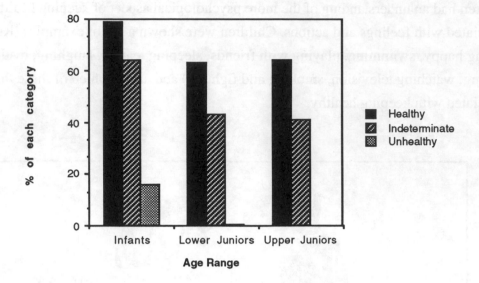

Fig 3.6: Table showing the % of all possible activities in each category selected as 'healthy' by each age range.

The limited evidence presented here suggests that children have already internalised an everyday understanding of what constitutes a healthy food and what are healthy actions/feelings by the time they have entered school. Hence the intervention was not expected to show significant changes in their responses. Whilst further work and reinforcement may be valuable for some children, progress from this baseline requires a knowledge of the body and its parts to understand the effect of common illnesses/diseases and chemicals on them. In addition, the different functions of food and their effect on the body have to be understood in order to address the concept of a balanced diet.

What knowledge of the body did children have?

This proved to be a truly fascinating area of the research. Although substantive work has been undertaken by other researchers[4], it has been done by those working in the domain of nursing and psychology and not in education. Hence the research provided an opportunity to explore the area and add to the knowledge of children's understanding of their bodies.

---

4   See Gellert, E. Children's Conception of the content and functions of the human body.
         Genetic Psychology Monographs, 1962, 65, 291-411 and
         Carey, S. *Conceptual Development in Childhood.* Cambridge.MIT Press. 1985

Children were asked to draw on an outline of the body what they thought was in their own bodies. Two responses are shown in the Fig 3.7 & 3.8.

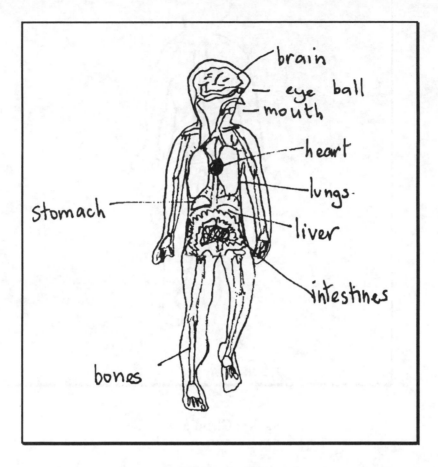

Fig 3.7: Richard[5]-age 10

The interesting aspect of this type of question is that it does not suffer from the weakness of interviews where the child may attempt to articulate a verbal response to an intuitive or incomplete view which may be taken too literally by the interviewer. The majority of children enjoy drawing and will attempt some representation of what they know. Where the nature of the representation was not clear, the children were asked to say what they thought it represented.

Fig 3.7 & 3.8 show two contrasting answers from children of the same age. The former shows a detailed biological knowledge with the organs drawn approximately to size and placed in the correct position of the body. Very few children were capable of providing such an answer. In contrast, Fig 3.8 shows a very limited understanding with only two parts drawn, neither of which is the correct shape or correctly placed. These drawings are shown to exemplify the range of answers which can be produced by upper junior children.

---

5   The annotations to this diagram are those of the interviewer.

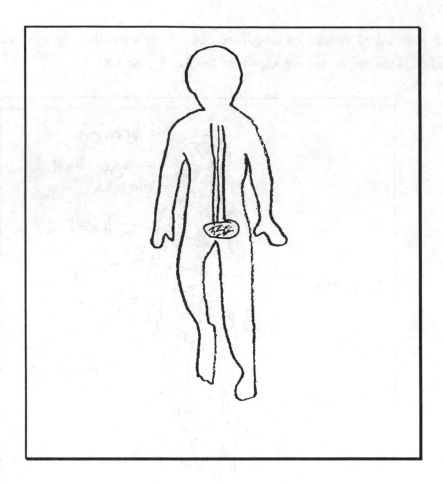

Fig 3.8: Claudia - age 10

Predominantly, children provided answers which incorporated the heart, bones, stomach and brain. Fig 3.9 shows that significant percentages of each age group mentionied these organs/parts of the body. The number mentioning the brain grew rapidly across the age range, and for infants the most commonly mentioned parts were bones. The decline in the frequency with which bones were mentioned by lower and upper juniors is inexplicable. When children are asked to *name* the parts of the body, other studies have found that blood is mentioned a large number of times as well.

On average, infants drew 3.4 organs or parts of the body which had increased to 5.0 for upper juniors. A fairly typical answer from a younger child is shown in Fig. 3.10. The increase in the number of parts drawn or mentioned supports the idea that children's biological knowledge naturally develops over this period of growth.

Fig 3.8 also reveals one of the other problems for children which is that many are not aware of the correct size and/or location of the organs. Fig 3.11 shows an example of a child's conception of their lungs.

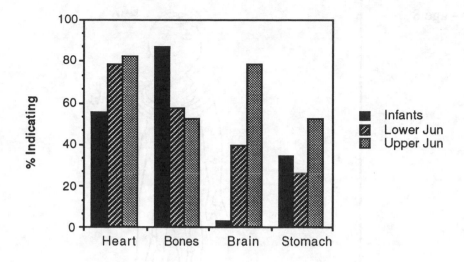

Fig 3.9: Most often mentioned organs of the body and % indicating

Such errors are not surprising since internal organs by their very nature are not visible or available for touch. Therefore it is difficult for the child to develop a knowledge of an object which can only be partially sensed. This research shows that children draw those organs or parts which are more easily sensed - the heart which beats, bones which can be felt and the brain because the capacity for self-conscious reflection and awareness has developed in children by this age. In general, organs such as kidneys, lungs, intestines are not sensed and not part of everyday language i.e. ' use you brain', 'my stomach aches', 'he's got no heart', which may provide one explanation for their lack of awareness of these organs.

Further evidence of a similar problem came from another question which asked children to add to an outline of the body to show where their heart was. Over two thirds of all the children drew the heart as a valentine shaped object (Fig 3.12). Similarly over two thirds of lower and upper juniors placed the heart on the left of the body rather than in a central location.

Infant drawings showed a wider variety of placements for its location which may indicate some uncertainty in their knowledge. Developing a knowledge of the heart and its position poses a particular problem for science educators because of the overwhelming number of everyday images which erroneously represent the heart as a valentine shaped object on the left of the body. One explanation for the data may be that infant children had not fully internalised this everyday image but this had certainly occurred for lower and upper junior children.

Fig 3.10:
Susan - age 8

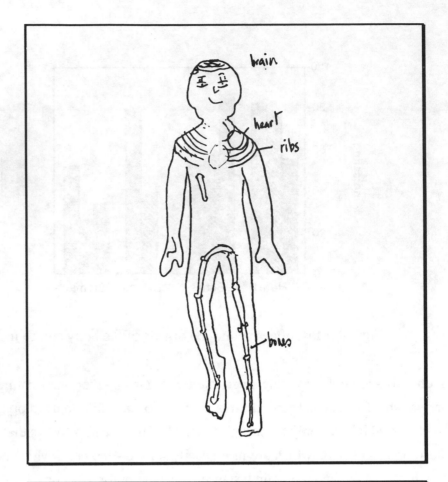

Fig 3.11:
Natalie - age
10.

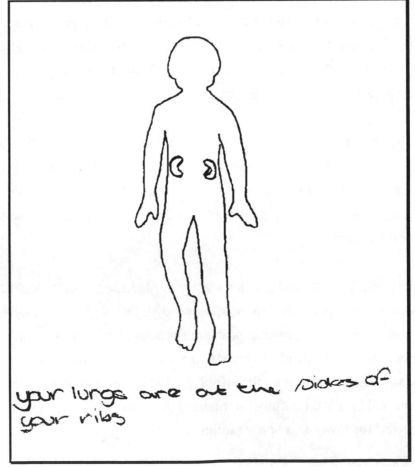

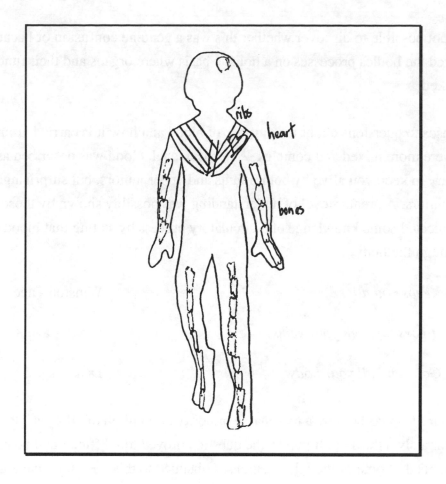

Fig 3.12: Clare - age 8

Further questions explored what these children thought the heart did and what blood was for. Typically infant children responded by indicating that the heart beats.

Beats and beats and beats and beats	Vandna-age 6
The heart beats	Tarana-age 5
Keeps you breathing and its flicking	Hamera-age 6

A greater depth of biological knowledge is shown in the understanding of older children, the majority of whom explained that the heart pumps blood.

> *Your heart beats every day and night and pumps blood into your veins. If it stops you will die*
>
> Kelly - age 9

A few children associated its function with respiration.

> *It helps you to breathe*   Lan - age 9

It was not possible to discover whether this was a genuine confusion or because they perceived the bodies processes on a holistic basis where organs and their function are interlinked.

Responses to questions about the purpose of blood and how it is carried around the body were more mixed and complex. At a basic level, blood was described as necessary to keep you alive by both infants and upper juniors, but surprisingly not by lower juniors. A greater level of understanding was possibly shown by those children who indicated some knowledge of a circulatory process by stating that blood moved or ran through the body.

> *Keeps you alive* — Winston - age 10
>
> *It goes through your veins* — Kevin - age 8
>
> *Go round all your body* — Dean - age 5

The latter idea was held by a reasonable minority of children of all ages. However, for approximately a third of all pupils, the question proved too difficult and no response was obtained. Some of the other responses obtained to this question give a glimpse of some of the ideas, some of which are quite logical, that children can hold about the purpose of blood.

> *It makes you stand up* — Anthony - age 8
>
> *Keeps your skin clean* — Susan - age 8
>
> *It lubricates the joints* — Andrew - age 10
>
> *It runs good food around the body after it has been digested.* — Edwin- age 9

The final response approximates most closely to the scientific view but was expressed rarely by children. Children's ideas about how blood is carried around the body, showed a range of thinking. There were a number of younger children who tended to think that blood moved itself or that body movement helped it to move.

> *When you walk and do things* — Dean - age 5
>
> *It moves itself* — Tumseela - age 5
>
> *It moves around when you wiggle* — Kathy - age 6

The latter notion carries with it the view of the body as an empty vessel around which blood sloshes. Many lower and upper junior children mentioned the veins in such responses and this shows a greater biological knowledge.

It goes through your veins   Dustin - age 6

Interestingly, the term 'arteries' was never mentioned by children and this would seem to be reflection of the lack of everyday use of this term.

The final question in this series simply asked children where in the body they thought they had muscles. The result of the elicitation are shown in Fig 3.13 which shows the two locations mentioned by more than 50% of all children, their arms and their legs. The next most-mentioned location chosen was fingers (Infants 21%, Lower and Upper Juniors 0%).

The trend of the intervention was to improve children's awareness of the numbers of parts in their body with muscles. However, this evidence again supports the notion that these children were only aware of those parts of their body which can be directly sensed or perceived i.e. muscles in the arms and legs.

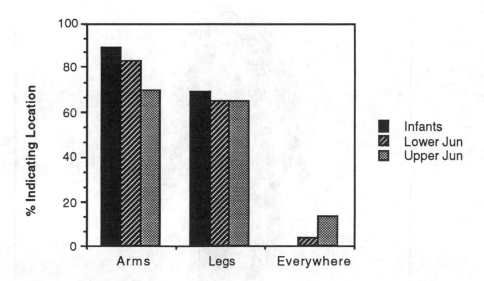

Fig 3.13: Percentage of children mentioning that muscles could be found in the arms, legs and everywhere in their body.

What processes are performed by components of the body?

The research chose to examine children's understanding of the process of digestion and respiration. The process of sexual reproduction was avoided because of the difficulties of conducting research in this area with young children, and the concepts of growth and excretion were explored through asking children whether they thought a range of objects were alive, not alive or once living.

To examine what children understood about the process of digestion, children were asked to add to an outline of the body to show what happened to food in their body. This question produced a wide range of responses which illustrated the range of children's understanding and the development of their biological knowledge. At the simplest level, children would simply draw a body cavity (Fig 3.14) containing untransformed food with no tube. Alternatively, food would be shown distributed through the human body (Fig 3.15).

Fig 3.14: Sashidaren - age 6

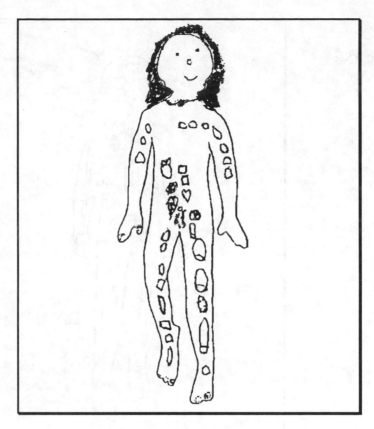

Fig 3.15 Tumseela-Age 5

As with all children's drawings, these and those that follow raise the question whether such drawings represent the limits of children's knowledge or alternatively, the limits of their representational capabilities. Firstly, it should be noted that only younger children produced drawings of this type and that the drawings show a lack of recognition of any physical connection between the mouth and the stomach or inside of the body. Secondly, some children who provided such drawings would qualify them with statements such as 'It (the food) goes into the blood. The blood goes everywhere.', which would suggest that they recognise that there was a process of at least partial transformation and that the drawings may not represent the limits of their understanding but the limits of their ability to represent food.

What these drawings lend support to is the view that an understanding of the process of digestion requires a comprehension that food can be transformed and broken down into its constituents. Till children understand this idea, the process of digestion will remain a mystery to them.

The next feature to emerge in children's responses was the tendency to draw two tubes from the mouth to the stomach. Fig 3.16 and Fig 3.17 show two good examples of such a drawings.

Fig 3.16:

William - age 9[6]

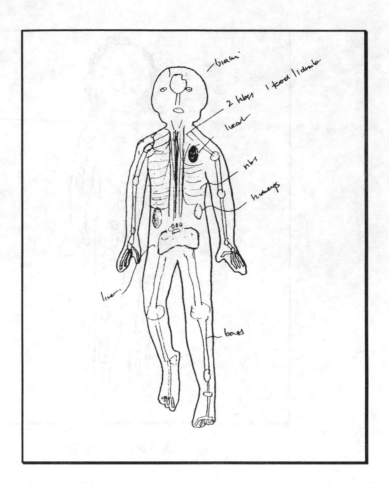

Fig 3.17:

David-age 8

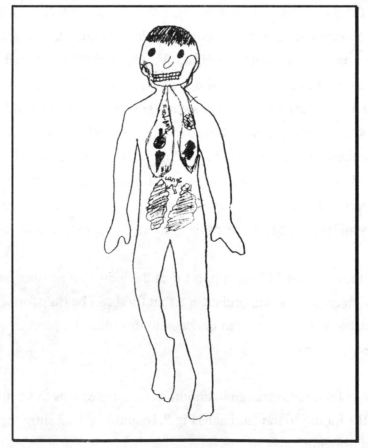

6   The annotations to this diagram are those of the interviewer.

SPACE Report   *Processes of Life*

In one sense, such drawings are a clear demonstration of children's logic in trying to reconcile their ideas to their observations. Waste products emerge from two different points in the body as liquid and solids. Differentiation clearly takes place and these drawings show a sensible attempt by children to explain their perceptions. It is also worth noting that everyday language i.e. 'it's gone down the wrong way' reinforces the concept of two tubes implying that there is more than one way for food or drink to pass through.

Progression towards a scientific understanding was shown by children whose answer only contained one tube. Fig 3.18 shows a good example. The stomach in such drawings was invariably placed in the centre of the abdomen and referred to generally as 'the belly' or 'tummy'.

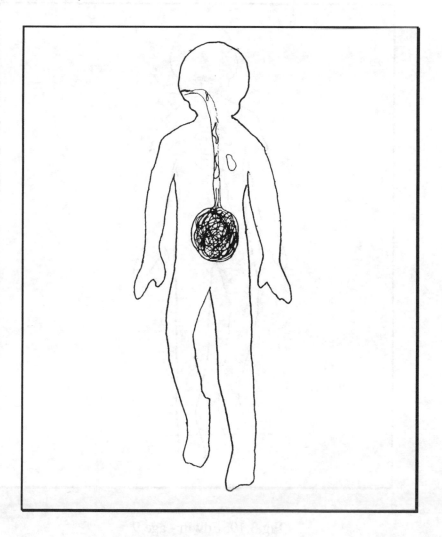

Fig 3.18 Chris - age 9

Such drawings lack any detail or understanding of what happens beyond the stomach. This is in fact the hardest aspect for most children. No infants indicated any aspect of the digestive tract beyond the stomach and only a minority of lower and upper juniors did so and an example is shown in Fig 3.19. This would indicate possibly that excretion is a relatively poorly understood process by children under 11. An alternative explanation is that eating and excretion are seen as two separate processes by children and not one continuous process.

The response Fig 3.19 represents a relatively sophisticated response in that the drawing shows a unitary digestive tract and locates the stomach in an approximately correct position. Only older children produced such drawings and this, coupled with the evidence of their greater knowledge of internal organs, lends support to the thesis that children's biological knowledge develops between the ages of 5 and 10.

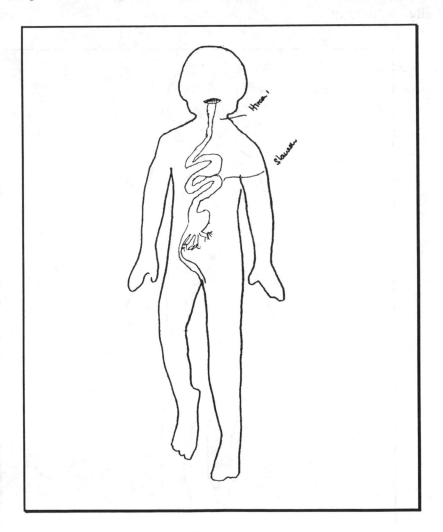

Fig 3.19: Edwin - age 9

What this data also supports is that the child's conception of the body is limited. For many children, it was restricted to that which is directly perceived or sensed. Knowledge that transcends such direct experience is only developed with difficulty over substantial periods of time and it is too easy for teachers to underestimate some of the difficulties children have in this domain.

The question 'Why do we need to eat?' was also used to explore children's understanding of the process of digestion and the purpose of food. The predominant response at all ages showed that the process of eating food was seen simply in everyday terms of its outcomes - eating food keeps you alive, enables you to stay healthy, to grow or get stronger. Essentially, these can be described as macroscopic, holistic responses and examples were:

If you didn't eat you would be skin and bones and die.   Winston: Age 10

So you can grow up and get fit   Steve: Age 10

To keep us alive   Clare: Age 8

To keep you healthy   Jennifer: Age 7

So you get stronger   Tumseela: Age 5

That means you can play when you eat your dinner.   Afsham: Age 6

There was no indication that any specific foods were seen solely as providing energy or assisting growth. Only two children gave any indication that food supplied chemicals that are essential for life by saying that food provided vitamins. This outcome is possibly not surprising. Whilst many children recognised that food could be broken down into smaller pieces, to understand that food contains different components which are absorbed into the bloodstream and distributed around the body, requires a particulate view of matter which is conceptually difficult for children of this age.

As a consequence, one of the activities of the intervention provided an opportunity for children to investigate foods and their categorisation - attempting to introduce the notion that some foods were best for body building, some for energy giving, some for keeping you healthy and whilst others made you fat. The development of this understanding is the essential precursor to a further understanding of food, the process of digestion and its outcomes.

Respiration

The research also attempted to explore children's understanding of respiration by asking children 'What happens to the air we breathe?' Answers to this question essentially fell into three categories. Predominantly, responses given were of an everyday nature.

It comes out	Kelly: Age 9
The air what you breathe in goes down to your belly.	Felicia: Age 10
It goes inside your tummy	Ettorino: Age 6
Keeps us alive and well	Lekan: Age 10

Such responses were restricted to clearly observable aspects of respiration and the overall consequences of breathing by contrast to dead or inanimate objects.

The second category of response was one which revealed some greater knowledge of the organs involved in respiration or of the process itself. Such knowledge is not self-evident and shows an improved level of understanding

It goes through your lungs	Rachel: Age 9
Goes in your lungs-goes into your heart from the lungs	Leon: Age 8

The final category of response which was very rare was one which showed a knowledge of gaseous exchange.

Air comes down and carbon dioxide comes out	Joanna: Age 11

It was found that there was a high correlation (0.92) between those children who mentioned gaseous exchange in their responses and those who showed lungs on their drawings of what is inside the human body. This would imply that an understanding of respiration is dependent on the development of children's biological knowledge.

A full understanding of respiration requires a knowledge of the organs involved, the role of breathing in performing gaseous exchange and the processes that occur within the body at the microscopic level to release energy from food. None of these children showed this level of understanding but the latter examples showed that these children had a greater knowledge of aspects of the whole process.

Application of processes of life to discriminating living and non-living objects.

Children were asked to say whether a list of ten objects (a plastic box, a small piece of rock, a spoon, a plant, an animal, an insect, an apple, a toy car and a seed) were living, once living or had never lived. This area of research has been a focus of attention for over sixty years[7] and used as a means of studying children's causal reasoning. However, the main interest in this work was to use such a question as a means of eliciting children's biological knowledge and not as a means of exploring the causal reasoning of children.

The range and diversity of children's response to this question provides a fascinating insight to children's thinking. There is not space to exemplify the range of children's reasoning but three such responses are offered as examples.

Object	Response	Reason
Plastic box	Never living	I don't know
A small rock	Once living	It was an animal once and it turned into a rock because my dad's friend has got thousands in the house.
A spoon	Never living	Cos metal is made
A plant	Living	Because it's growing
An animal	Living	It's made in an egg
An insect	Living	It's made just like other animals.
An apple	Once living	It was alive when it was growing on a tree
A toy car	Never living	Not sure
A seed	Once living	They came off another plant

Leah - Age 9.5

This response shows the child's use of a number of criteria. Objects were distinguished by the fact that they can grow, reproduce, are man-made and that they originated from living material. A slightly different response is shown next.

---

[7]   See Piaget, J. (1929) The Child's Conception of the World.New York: Harcourt Brace for the first work undertaken in this area.

Object	Response	Reason
Plastic box	Never living	It hasn't got a face
A small rock	Once living	Still got no face
A spoon	Never living	It's for putting food in your mouth
A plant	Living	It's growing and growing
An animal	Living	It's got a face
An insect	Living	They fly
An apple	Never living	You have to eat an apple
A toy car	Never living	It's got no face
A seed	Once living	You have to make a hole in it

Jermaine - age 5

Here the child's reasoning centrates on the external features and shows the repeated use of the criteria of whether it has a face or not. Other responses show the use of the criterion of movement and purpose. Finally the third example shows a response which was typical of many infant children.

Object	Response	Reason
Plastic box	Living	It's round and it's got a hole
A small rock	Never living	Because it hasn't got any holes.
A spoon	Never living	Because its only got a hole
A plant	Living	It's round and big.
An animal	Never Living	It's round and hasn't got any holes
An insect	Never Living	They haven't got any holes
An apple	Never Living	It hasn't got any holes
A toy car	Never living	Because you can't open the doors
A seed	Never Living	No holes in them

Arridet - age 6

This response show the child focusing on a single external feature and repeatedly using this criteria. The simplest explanation of this response would be that the child only recognises visible external features and attempts to use these as a criterion in responding. There is also the possibility that such a response represents the child's attempt to articulate an explanation for a concept that has only been intuitively recognised. Once the child has managed to state an answer for the first time, they continue with the consistent application of the same criterion and do not recognise the need for more thought about the response.

Clearly a scientific response would demand the application of a consistent set of criteria - that is whether the object showed any of the features of the process of life (movement, growth, reproduction, digestion, respiration, sensitivity and excretion). The extent to which these responses were used is shown in Fig 3.20

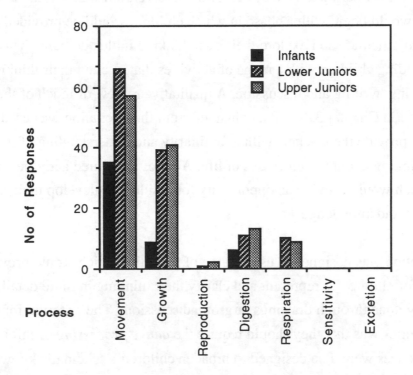

Fig 3.20: Chart showing the number of responses from each age grouping which mention specific processes of life.

The data show that the predominant process of life used as a criterion by children was that of movement and growth, and that such criteria were more extensively used by lower and upper junior children. Other processes of life were used as criterion in a very small number of instances out of a total possibility of 675 opportunities to apply criteria. The data does show that both lower and upper juniors mentioned reproduction, digestion and respiration more often than infants but even then, only in a very small number of instances. The fact that the question reveals that some lower and upper junior children knew of these processes would support the hypothesis that there is a natural development in children's biological knowledge during these years.

4: The Intervention Phase

The previous chapter provides some insight into the range of ideas about the processes of life held by young children. Whilst this qualitative picture is valuable in providing an insight into children's biological knowledge and understanding, the aim of this research was to attempt to extend previous work by devising a set of intervention activities which could be used by teachers to develop children's thinking and biological knowledge.

The rationale that underpinned the design of the intervention was that the teaching and learning would begin with a phase in which children would be provided with an opportunity to articulate and explore their own thinking in this domain. This was done by providing children with a range of activities that elicited their thinking through drawing, writing and discussion. A qualitative review of much of the data has been presented in Chapter 3. The data obtained from the elicitation was used informally to provide the teachers with a familiarity and understanding of their children's thinking about the processes of life. A set of structured activities was then provided which would provide an opportunity for children to develop their understanding and knowledge.

This intervention was designed to use a range of activities which would provide an opportunity for children to represent and clarify their thinking in more detail. This was generally done through drawings or group discussion. The criterion for selection of these activities was that they should require the *active processing* of information. These experiences were also designed to broaden children's schematic knowledge, extend their vocabulary and where appropriate, generate a conflict between their thinking and experience which would lead to a re-evaluation of their ideas.

The selection and design of the activities for the intervention was influenced by three factors

    (a)   A preliminary analysis of the data

    (b)   A set of ideas defined by the 'scientific' understanding (Chapter 2 - 'Defining the Processes of Life') which would assist a child in developing an understanding of the scientific world view.

    (c)   The teacher's contributions and ideas.

The elicitation gave a broad picture of the level of children's knowledge and understanding in this domain. Essentially, many children's knowledge of the body and of its processes was limited to external features and there was therefore a need to provide opportunities to develop their understanding of the internal components and their function. Unlike some other aspects of science e.g. electricity and light, such knowledge cannot be shown or developed through empirical investigations which are a feature of much physical science. Hence, the intervention used a range of broad strategies which were available for teachers to use whenever they judged appropriate. These can be described as a) sorting activities b) discussion activities c) modelling/making activities and d) investigations.

Sorting activities.

These activities require the active processing of information by children. Typically they would be provided with a number of cards. Each card would have a food on it and the children were asked to sort the foods into groups. Invariably, to start with children often sorted them into 'foods they liked' and 'foods they did not like'. The role of the teacher was then to encourage children to devise other ways of grouping the foods. One suggested activity for older children, was that food labels were cut off packets and then the labels sorted by the categories of information on the labels to encourage children to explore the meanings of the data presented in food labelling. However, teachers were always asked to provide children with ample opportunity to explore their own approaches to categorisation.

Another use of sorting was to provide children with a set of cards, each with a part of the body the written on it e.g. ear, mouth, lungs and another set of cards with the function on e.g. for hearing, for chewing food, for taking in oxygen. Children were then asked to match the names on the cards with their functions as a group activity.

A third approach was based around the use of simple classification activities. Sets of objects were provided and children asked to sort them into living and non-living. Children used their own criteria to start with but each time they used one criterion, they were then asked to think of another. Older children were encouraged to use more complex forms of classification to derive a wider range of groups e.g. Does it move? Does it live in water? and they were encouraged to use a variety of computer programs which enable such classification.

Discussion Activities

Many of the sorting activities discussed previously were undertaken by groups and hence required discussion and communication between peers which encouraged both articulation of their own thinking and the exchange of ideas. Wherever possible, activities were used that encouraged the use of this technique.

For instance, children were asked to discuss in groups such questions as 'How do healthy people look?' 'What do healthy people do?' and to produce a message for not so healthy people. In another activity, children were asked to draw a picture of what they thought was inside the body and then discuss each others' pictures and produce a group picture which they felt was most nearly correct. Further details of such activities can be found in Appendix 3.

Modelling/Making Activities.

Models provide a tangible and concrete experience of objects which are not readily open to inspection such as the inside of the body. In one activity, children were asked to feel all their bones and then compare their experience with the representation shown on a cut out model of a skeleton. For older children, another cut-out was used where children were asked to place parts of the body on a large cut-out. Making posters of 'things that make us feel good' and 'things that make us feel bad' or large posters of 'energy foods' and 'body building foods' was also encouraged as a active means of enabling children to share and discuss their thinking.

Investigations

The general principle underpinning the SPACE programme was that children should be provided with an opportunity to design their own investigations with whatever equipment was easily available. In this domain, the range and scope for investigations is limited. However, appropriate investigations were suggested to teachers in the event of the children failing to devise an appropriate investigation or to supplement the activities devised by the children. Simple stethoscopes were made with plastic cups and rubber tubes. Pulses were felt and timed and children were asked to investigate the location of muscles in their own body.

Teachers were provided with a list of a possible activities (Appendix 3). These were seen as essentially a resource which could be used with children, as and when it was appropriate to the knowledge and understanding. The intention was not to provide a prescriptive set of experiences but simply a set of activities that were available for use. Teachers were encouraged to invite children to devise their own methods of testing their thinking. This was not always possible and, to broaden the experience of children, some of the activities suggested were used by most teachers.

General Issues

Although the data collection was undertaken by the researchers, the intervention work was undertaken by the teachers. During this phase, the researchers made regular visits to the schools to support the teachers and to share with them the data collected after the preliminary elicitation. Teachers who undertook to work on this project were given briefings about the nature of the approach and the need to elicit children's understandings of the particular concept of interest before commencing teaching. Moreover, it was emphasised to the teachers that the nature of the individual child's understanding should be the basis for determining the intervention work. That is, that they should attempt to ascertain what the child already knew before determining the strategy for teaching and learning. Sharing the data gathered from the elicitation with the individual teachers was one way of enabling this process and was undertaken in all instances. In addition, teachers were encouraged to undertake similar activities in the classroom to provide more insight for themselves.

No attempt was made to ensure consistency of experience between one classroom and the other. Variation is inevitable and a reflection of the normal classroom realities. Teachers were briefed about the general approach to the intervention and the strategies to adopt and asked to offer children a wide variety of experiences and opportunities to investigate topics of interest. The briefing document which was the basis for discussion with the meetings with teachers stated:

> ' We suggest that you carry out at least one activity from each section[1] and then as many others as you are able to. We would like you to keep a log of all the activities which you try, noting how successful you felt they were, how the children responded and how you were able to build on the activities. It would

---

[1]   See Appendix 3

also be helpful if you could record interesting comments made by the children and save copies of interesting/typical work.

It is important that most of the investigations stem from the children's ideas and are not presented to them in isolation. They may need to talk, write about or draw their ideas before embarking on an activity. Wherever possible, the activities should be initiated by the children in response to open-ended questions e.g. "How could we find out about....." or "How could we find if that it is true?"

Although children may wish to consult secondary sources for further information, this should be done in conjunction with practical activities, not "Let's look it up in a book!" first. Equipment available in the classroom for children to plan their own investigations should be a useful starting point for many of the activities.

<u>Exploration</u>

A table or corner could be set up in the classroom with the following equipment available:- a stethoscope (possibly homemade), a model skeleton, a forehead thermometer, balloons, bones, hand lenses, mirrors, string, timers, books about the body etc. Children could be given specific times to use the equipment so that they are able to devise their own investigations. They should also be given the opportunity to share their ideas with the rest of the class.

An area could be set up with some of the questions which the children were asked during the elicitation phase e.g. a collection of objects could be available to them to classify. They could be asked to sort them into three sets: - one for living (green), one for never living (red) and one for once-living things (blue). When they had sorted them they could enter their result onto a graph using the same colours to colour each square. They could use this data base for discussion and deciding which criteria they would use to decide if an object was alive or not.'

Consequently, the data obtained from this study cannot be used to judge the validity of any one activity but merely provide an analysis of the potential developments in children's thinking from exposure to a range of experiences which embody the broad strategies outlined here.

5:The Effects of the Intervention

This chapter provides a full analysis of the data gathered pre and post-intervention. The data were gathered using a mixture of written questions and interviews which are provided in Appendix 2. Groups of children, of about 4 or 5 in number, were asked to write their answers to questions 1-7 and all the questions that required any drawing e.g. a drawing of what is inside your body or a drawing of four things that they do to keep healthy. Responses obtained were then discussed with the children in individual interviews to obtain further clarification of their meaning and the children's answers annotated by the interviewer. A set of 9 objects/drawings were then presented individually to the child and the question asked "Is this .... living, once living or never living?". The child's responses were then recorded by the interviewer.

Data were gathered in two phases, an elicitation phase prior to the intervention and a further elicitation phase after the intervention. These two phases were generally separated by a period of 6-8 weeks as the intervention work was undertaken over a 'half-term' period. The questions used in both phases were identical and had been designed to explore children's understanding using a wide range of activities which gave children the opportunity to write, talk and draw as a means of expression. In the case of infant children, all the data collection was undertaken by a process of individual interview because of their generally limited capabilities in expressing themselves in writing. All the data were gathered by the full-time project officer and a part-time researcher.

The data collected explored five themes of children's understanding which were identified in the English & Welsh national curriculum[1]. These were considered to be:

a.   *What choices and actions are required for healthy living?* These issues were explored by the use of Q1 and 2 (Appendix 2); asking the children to draw four things to do with healthy living and to draw a healthy meal and an unhealthy meal.

b.   *What processes are performed by components of the body?* This understanding was explored by Q 3, 4, 5, 6; asking what the function of the heart was and to add to an outline drawing to show what happens to food and drink inside the body.

---

[1]   Department of Education & Science. (1989) Science in the National Curriculum. HMSO. London

c.   *What knowledge of the body did children have?* Question 7 and the questions asking children to draw 'where the heart was' and 'what else is inside your body' were designed to explore children's factual knowledge and awareness of of their own bodies.

d.   *What weighting or association is given to the processes of life in the child's concept of 'living thing'?* This question was explored by the use of a set of objects which where shown to the child who was asked to state whether the object was 'living', 'once living' or 'never living'.

e.   *What was the child's knowledge of plants and their parts?* Only one question was used to explore this aspect of their knowledge where children were asked to label a drawing. The limited exploration of this aspect was in part a reflection of the apparent emphasis within the national curriculum and in part a reflection of the difficulty of exploring understanding in this domain and the priorities of the research which placed more emphasis on children's understanding of their own bodies and their maintenance.

Questions (a) to (d) were addressed through multiple items which enabled the possibility of exploring the consistency or lack of it which children used in their answers. (See Appendix 2 for the full set of questions).

The data presented are those obtained from children who were present on all three occasions i.e. the elicitation, the intervention and the second elicitation. Full sets of data were obtained from 75 children in total. This consisted of 23 Upper Juniors in year 5 & 6 of their education, 23 Lower Juniors, in year 3 & 4 of their education and 29 infants in year 1 & 2 of their education. Sample sizes for the different age groups varied but each sample was taken from a minimum of two schools. One of the difficulties that emerges in research of this nature is the considerable diminution in the sample caused by the absence of children in one or more of these phases.

The methodology used in analysis of the data was firstly a simple categorisation of the answers and a frequency count. Some of these categorisations were based on the researcher's understanding of commonly accepted understandings i.e. healthy and non-healthy foods. Other categorisations were based on an empirical approach to the data from the responses provided by children.

For those data where there were two or more aspects to the response i.e. the nature and function of the blood, the data were analysed using systemic networks[2]. These

---

2   Bliss J., Ogborn J. & Monk M., (1983) Qualitative Data Analysis. Croom Helm

networks allow for several parallel aspects of individual responses to be viewed in conjunction and present a more holistic impression of the concept that children may be using to answer elicitation questions on the same topic.

Data are conjoined through the use of one of two devices, called a 'bra' or a 'bar' respectively, for which the symbolic representations are shown below (Fig 5.1 & Fig 5.2). Fig 5.1 shows part of the network to code children's drawings of the heart. The child either draws a heart which is too large, approximately the correct size or too small, but obviously cannot give more than one of these responses. Each of these responses is called a 'terminal' and counts can be made of the number of responses classified by each terminal.

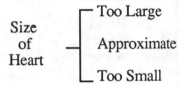

Size of Heart
- Too Large
- Approximate
- Too Small

Fig 5.1 An example of a 'bar' used in systemic networks

A 'bra' is the converse in that the categories are inclusive and the response of the child may be in one or more of the categories. Hence children's responses about their knowledge of the heart may contain aspects about the location, shape, size and function. All of these responses are counted at the terminals.

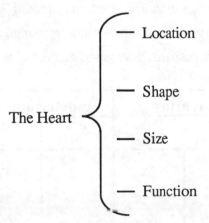

The Heart
- Location
- Shape
- Size
- Function

Fig 5.2: An example of a 'bra' used in systemic networks

In logic terms a 'bar' is exclusive whereas a 'bra' is inclusive. The counts made at the terminals were analysed using chi-squares to see if there were significant changes as a consequence of the intervention.

What choices and actions are required for healthy living?

There were four items which explored children's understanding of this issue. Question 1 presented a list of items and children were asked to ring those ones which they considered to be 'healthy foods'. Clearly the notion of what constitutes a 'healthy food' is culturally relative decision in that a starving person would consider any food 'healthy'. However, for the purposes of this research, it was agreed by the researchers that, within the the context of children's everyday experience of our British culture, the foods listed could be categorised into 3 groups, 'healthy', 'indeterminate' and 'unhealthy'.

Healthy	Indeterminate	Unhealthy
Lettuce	Bread	Chips
Apples	Meat	Coke
Rice	Potatoes	Burger
Juice		Biscuits
Brown Bread		Sweets
		Crisps

Table 5.1: Table showing categorisation scheme used for analysis of Q1

Children's responses were then analysed and categorised on the basis of this schema. For each child, the number of 'healthy', 'unhealthy' and 'indeterminate' responses were counted. These were then totalled for the whole group and the table 5.2 shows the percentage of all total possible responses indicated as 'healthy' by each group of children. Thus prior to the elicitation upper juniors marked 75% of all the possible healthy responses as 'healthy', 62% of all the indeterminate responses as 'healthy' and only 3% of all the unhealthy responses were indicated as 'healthy'.

Age Grouping	Healthy		Indeterminate		Unhealthy	
	Pre %	Post %	Pre %	Post %	Pre %	Post %
Upper Juniors (n=23)	75	87	62	47	3	2
Lower Juniors (n=23)	69	66	65	69	11	4
Infants (n=29)	77	75	79	78	34	22

Table 5.2: Total % of all the foods in each category indicated as being a 'healthy food' by each age group

The data show that upper and lower juniors are clearly knowledgeable about which foods are commonly considered to be 'healthy' and 'unhealthy' since only a very small percentage of the unhealthy foods were marked 'healthy'. The data show the effects of the intervention was limited and only the upper juniors showed evidence of a more refined concept of what constituted a healthy food, by indicating that a greater percentage (87%) of the foods categorised as healthy were so, and a smaller percentage (47%) of the indeterminate foods were 'healthy'. An analysis of this table using chi-square tests shows that from pre- to post-elicitation, this change was significant (p<.05).

The data for infant children would indicate that the concepts of 'healthy' and unhealthy foods are not so well formed as approximately one third (34%) of the unhealthy foods were indicated as being 'healthy'. The difference between infants and lower juniors was highly significant (p < .001) both pre- and post-elicitation. Since infants marked many more of the unhealthy foods than lower juniors as 'healthy', this would possibly suggest that during this period there emerges the recognition that 'you eat to live' is too simplistic and that some foods are essential or necessary for life, whilst others are unnecessary or can even have a negative effect.

For the 'indeterminate' category, a similar, but substantially less-significant effect (p<.05) was found between infants and lower juniors (pre-elicitation only). A variation of greater significance (p<.01) was found post-elicitation between lower juniors and upper juniors for this category.

The general approach in designing intervention strategies was to devise activities that required *active processing* of information requiring children to articulate their concept to others or themselves. In this case, one of the activities recommended was a food sorting activity where children were asked to sort foods (or cards carrying the names of foods), initially using their own categories and then those suggested by the teacher. Whilst the evidence is that such activities did not achieve their purpose of encouraging children to discuss types and categories of foods, initial categories tended to be 'foods they like', 'foods they do not like' and perhaps it is this categorisation that is remembered more than the categories of energy giving foods, body-building foods and healthy foods which were suggested by the teacher. Secondly, this would be the first time that children had probably met such categorisations and their meanings may not have become self-evident from a limited encounter. Whilst the evidence from this research was that such a strategy had a limited effect, it is notable that children can identify healthy foods in over two-thirds of all the instances presented to them. This inevitably begs the question-why, if children are capable of recognising healthy foods, are they so reluctant to eat them?

The second question attempted to examine the wider implications of keeping healthy by presenting children with a selection of thirteen activities and asking them to ring those which are 'to do with keeping healthy'. These were deliberately chosen to represent a very broad spectrum of activities to fully explore the nature and extent of children's understanding. Once again for the purposes of the research, categorisation of the responses was undertaken using the following framework which was agreed by the researchers.

Healthy	**Intermediate**	**Unhealthy**
Running Sleeping Feeling Happy Swimming Playing with Friends	Eating Arguing Laughing Reading	Watching Television Smoking Fighting

Table 5.3: Table showing categorisation scheme used for analysis of Q2

Table 5.4 beneath summarises the data showing the total percentage of all responses for each category indicated as being 'healthy' by each age group i.e 64% of all the possible healthy responses were marked healthy by the Upper Juniors

Age Grouping	**Healthy**		**Indeterminate**		**Unhealthy**	
	Pre	Post	Pre	Post	Pre	Post
	%	%	%	%	%	%
Upper Juniors (n=23)	64	71	41	42	0	4
Lower Juniors (n=23)	63	71	43	53	1	5
Infants (n=29)	79	77	64	62	16	14

Table 5.4: Total % of all possible responses in each category (healthy, indeterminate and unhealthy) indicated as being 'to do with keeping healthy' by each age group

Undoubtedly the most remarkable feature of this table is the consistency of the responses pre- and post-elicitation. There were no significant changes for any of the categories and clearly the intervention has failed to alter any notions children may have had of what constitutes a healthy activity. What the data do show is that children are aware, from an early age, that the activities of smoking, fighting and watching television can not be considered healthy activities as very low percentages of children across all age ranges marked these categories of response. Conversely two thirds to three quarters of all the choices categorised healthy were correctly indicated as 'healthy' with little variation between the age groups.

The only significant (p<.01) variation found in response was between the infants and the other two age groups prior to the intervention. They correctly indicated more of the healthy responses but also marked more of the 'unhealthy' responses as being 'to do with keeping healthy'. A possible explanation is that the significance can be explained by a tendency of infants to ring any response when unsure about the healthiness of the suggested activity in response to the question. Thus the data would not support the inference that they have a better understanding and knowledge of what constitutes a healthy activity.

The third question to address this matter was the one which asked children 'to draw four things which are to do with keeping healthy.' Children's drawings fell into two categories which were predominantly used and a range of minor categories which were more infrequently shown. The major categories were considered to be 'food and drink' and 'exercise and sleeping'. Minor categories were 'drugs', 'smoking' , 'vitamins' and a wide variety of other activities associated with keeping healthy. Food and drink was divided into three groupings of 'healthy', 'indeterminate' and 'unhealthy' following the definition shown in Table 5.3.

The data for this question were collected by ticking a category if one or more drawings judged appropriate to the category were present and the results are shown in table 5.5. Children who gave many drawings in one category are only counted once on this table whilst children who gave drawings in more than one category are counted twice or more. This is because the table attempts to reflect the breadth of the response understanding.

The responses to this question are similar to those discussed previously. There is little variation between any of the age groups either before or after the elicitation. The only significant variation (p<.01) was between infants and the other two groups. The latter's responses contained more diagrams indicating that exercise and sleeping were activities that kept them healthy. This feature of the data was still present after the intervention.

Category	Infants (n=29)		Lower Juniors (n=23)		Upper Juniors (n=23)	
	Pre	Post	Pre	Post	Pre	Post
Food						
- Healthy	23	23	22	14	20	18
- Indeterminate	0	1	-	2	-	1
- Unhealthy	3	5	-	-	-	-
Exercise & Sleeping	4	7	14	20	13	19
Drugs/Medicine	1	-	-	-	1	6
Smoking	-	-	-	-	1	4
Vitamins	-	-	-	-	3	-
Other						
- drink	-	-	6	2	2	-
- brush teeth	-	-	4	2	-	-
- washing	-	-	-	3	-	-
- watching TV	-	6	2	-	-	-
- going to toilet	1	-				
- playing	-	1				
Unclassifiable	3	-	-	-	0	0

Table 5.5: Total numbers of children giving drawings in each category

The intervention had only one significant effect (p<.05) in increasing the number of lower juniors who indicated exercise and sleeping as being an activity to do with keeping healthy. The numbers for the variety of 'other' activities are too small to assess their significance. The picture that emerges is one that is consistent with the answers to previous questions with little variation of response as a consequence of the intervention.

The last approach that was used to explore children's understanding of the choices and actions necessary for healthy living, was to ask children to draw a 'healthy meal' and an 'unhealthy meal'. Children were given a sheet of paper with an outline of two plates on it. One was labelled 'healthy' meal and the other labelled 'unhealthy' meal. Data were then collected of the types of food indicated. In all, 31 foods were drawn or mentioned by children. These were vegetables, carrots, peas, fish, tomatoes, potato, meat, lettuce, chips, fruit, hamburgers, bread, eggs, drink, rice, spaghetti, bacon, beans, brown bread, cake/biscuits, cheese, chicken, cod liver oil, cornflakes, milk, orange juice, sausages. The full data for the responses are given in Appendix A4. Table

5.6 beneath shows the foods mentioned by more than 25% of the children for each group in either the pre- or post-elicitation.

With the exception of the infants, these tables reveal that children have a clearly defined notion of what constitutes a 'healthy' meal. Carrots, peas and vegetables were the foods that predominate in the thinking of upper and lower juniors and it was notable that anecdotal evidence would suggest that it is these foods which are generally not appreciated by children. Whilst there was some variation between pre- and post-elicitation and some changes were significant, it would be difficult to ascribe a causal mechanism to the change in the choice of one 'healthy' food compared to another. Of more significance was that for the both lower juniors and upper junior, 5 out of the 6 foods are mentioned by more than 25% of all pupils both pre and post-elicitation.

Upper Juniors:Table 5.6a

Pre		Post	
Carrots	78%	Carrots	57%
Veg	61%	Peas	48%
Lettuce	52%	Lettuce	43%
Fish	35%	Rice	39%
Peas	26%	Meat	30%
Tomatoes	26%	Veg	26%
Meat	17%	Tomatoes	17%

Lower Juniors:Table 5.6b

Pre		Post	
Veg	74%	Carrots	78%
Carrots	70%	Peas	70%
Peas	65%	Veg	39%
Fish	48%	Fish	39%
Tomatoes	26%	Potatoes	30%
Potatoes	22%	Tomatoes	22%

Infants:Table 5.6c

Pre		Post	
Chips	41%	Fish	45%
Bread	41%	Fruit	31%
Fish	34%	Vegetables	28%
Fruit	34%	Peas	21%
Peas	28%	Bread	14%
Vegetables	17%	Chips	14%

Table 5.6 a, b & c showing the principal foods indicated as being 'healthy' and the % of each group doing so.

These data would suggest that children of this age had developed a well-defined set of criteria of what constituted a healthy food prior to the intervention. The research unfortunately did not explore these criteria and their nature. Such a line of enquiry would have value in exposing whether the nature of children's knowledge had an explicit rationale or was based on intuition and everyday reinforcement from the home.

For the infants, 3 out of the 6 most-often mentioned foods are identical before and after the intervention. More interesting is that only two of their foods, fish and peas, were mentioned by upper and lower juniors. Children in the latter two groups were significantly different in that they did not mention bread and chips so frequently. From a health education perspective, this clearly constitutes an improvement in children's understanding and is evidence of the value of such intervention work with children of this age. Perhaps surprisingly, infants were the only group to regularly mention fruit as a 'healthy' food.

A similar analysis was performed for the responses to the other half of the question which asked children to draw what they considered to be an 'unhealthy' meal.

Upper Juniors:Table 5.7a

Pre		Post	
Chips	83%	Chips	100%
Hamburgers	48%	Hamburgers	78%
Eggs	43%	Sausage	35%
Sausage	26%	Cake	26%
Cake	4%	Eggs	9%

Lower Juniors:Table 5.7b

Pre		Post	
Chips	83%	Chips	78%
Hamburgers	57%	Sausage	48%
Sausage	26%	Hamburger	43%
Eggs	26%	Eggs	30%

Infants:Table 5.7c

Pre		Post	
Sweets	79%	Sweets	55%
Eggs	34%	Chips	31%
Cake	34%	Eggs	7%
Chips	28%	Cake	7%

Table 5.7 a, b & c showing the principal foods indicated as being 'unhealthy' by children and the % indicating so.

Again, the notable feature of this data was the distinction between the infant group and the other two groups. Upper and Lower Juniors consistently indicated the same foods i.e. chips, hamburgers and/or sausages as being 'unhealthy', whereas the food which featured predominantly for infants was sweets. Chips only appeared as an item mentioned by a substantive number after the intervention. Two points can be made about this data. Firstly that these infants concept of a meal permitted the inclusion of sweets and that for them there was no demarcation between snacks and meals. More importantly though, it is clear that whilst the infants had a notion of what was unhealthy, their concept of this was different from that of lower and upper juniors. This would suggest that over this period, concepts of food and their value are in some process of development. Given that children appear to clearly appreciate what was unhealthy food and yet, continue to eat it, this phase of development would possibly be the appropriate moment to intervene and develop a better scientific understanding and more sensible attitude to food. Unfortunately the evidence gathered from this study does not sustain such a hypothesis but such a change may possibly not have occurred because of limited treatment of the issue by the intervention.

In summary then, the research provides a clear picture that these children from the age of 7 up were clearly aware of the role of exercise and the choice of food in sustaining a healthy lifestyle. In many areas, there was evidence to suggest that children had some understanding of such concepts prior to the intervention. This would in part, explain the lack of significance in any of the findings since the knowledge explored was already internalised by children. The origins of this knowledge were not explored but given that these ideas are a regular theme of much advertising, it is not unreasonable to suggest that much informal education occurs through the media.

What knowledge of the body did children have?

Children's understanding of the processes of life will be limited by their biological knowledge. For example, a child who does not understand that the body contains lungs made of a spongy tissue which enables the interchange of gases with the blood is unlikely to see respiration as anything more than the act of breathing. Therefore science education will need to develop an awareness in children of a range of internal organs and their function at an early stage to improve their understanding of these processes.

In this research, children's biological knowledge was explored through three questions:- one which looked at the range of locations in which children could identify muscles, one which looked at children's idea of the location of the heart and another which asked children to add to an outline of the body, what they thought was inside.

The first question asked 'Where in your body are muscles?' and the data obtained are as shown in Table 5.8.

	Infants		Lower Juniors		Upper Juniors	
	Pre %	*Post %*	*Pre %*	*Post %*	*Pre %*	*Post %*
Arms	86	79	83	78	70	70
Legs	69	59	65	74	65	74
Fingers	21	0	0	26	0	43
Feet	3	3	0	22	0	22
Neck	7	0	13	17	0	13
Belly	10	0	0	13	17	13
Toes	0	0	0	0	0	17
Jaw	0	0	0	4	4	4
Wrist	0	0	0	4	13	0
Chin	3	0	0	0	0	0
Elbow	3	0	0	9	4	4
Heart	3	0	0	9	0	0
Body	7	0	0	0	0	0
Shoulders	3	7	0	13	4	9
Back	0	0	4	9	13	9
Chest	0	0	0	4	4	13
Face	0	0	0	4	4	0
Everywhere	0	24	4	9	13	35
Other	3	3	4	9	9	17
No response	3	3	9	0	0	0

Table 5.8: Percentage of children mentioning the specific location of muscles in the body.

The picture that emerges is one where the majority of children perceived muscles as being in arms and legs but only a small minority of responses indicated that muscles could be in other parts of the body. The general trend of the intervention was to improve the awareness of muscles in other parts of the body with a significant ($p<0.05$) increase of the number of infants who stated that muscles were to be found everywhere in the body. This was accompanied by a significant decrease in the number of infants indicating that they were to be found in their fingers. Otherwise none of the changes are significant.

The data clearly define the limits of these children's knowledge and reinforce the notion that these children's knowledge of biology was limited to those aspects of the body for which there is an easily accessible direct experience. Muscles in the arms and legs can easily be sensed and felt, hence the readiness to state that this is where muscles are found. Muscles in other parts of the body are not so self-evident and their relation to movement was not appreciated by many of these children.

The next question to explore children's biological knowledge was one which asked children to draw (on an outline of the body) a picture of where there heart is. This question was chosen as it was felt that the heart is one internal organ which children are familiar with from a young age. Hence it was thought to be of interest to see what conception they held of this organ, its function and its location. The function was considered by a separate question which asked 'What does your heart do?. The data were tabulated in a network (Fig 5.3) in order to explore the relationships that may have existed in children's mind between the function, location and size.

Examining the data several clear points emerge. Firstly the overwhelming majority of children initially thought that the heart has a traditional valentine shape. The percentages holding this view are shown in Table 5.9

	Infants		Lower Juniors		Upper Juniors	
	Pre	Post	Pre	Post	Pre	Post
Valentine Shape	83	80	74	48	65	35

Table 5.9: Percentage of children holding particular conceptions of the shape of the heart.

However, the effect of the intervention was to diminish the number of lower and upper juniors holding this idea to a minority. Whilst the the size of the change for lower juniors approaches significance, it was only significant ($p < 0.05$) with the upper juniors. The other positive effect of the intervention, also significant ($p < 0.05$), was the increase in the number of lower juniors who indicated that the heart pumps blood. Whilst it is argued that this represents an improvement in these children's understanding of this organ, the research did not explore their conception of the circulatory system and there is evidence that many children perceive the heart working in an open or partially open circulatory system. That is some blood remains in the veins and some blood leaves to bathe the cells.[3]

---

[3]   Mintzes, J.J, Trowbridge, J.E, Arnaudin, M.W. & Wandersee, J.H., (1991). Children's Biology: Studies in Conceptual Development in the Life Sciences in Glynn, S.M., Yeany, R.H. & Britton, B.K (Eds) *The Psychology of Learning Science*, New Jersey: Lawrence Erlbaum

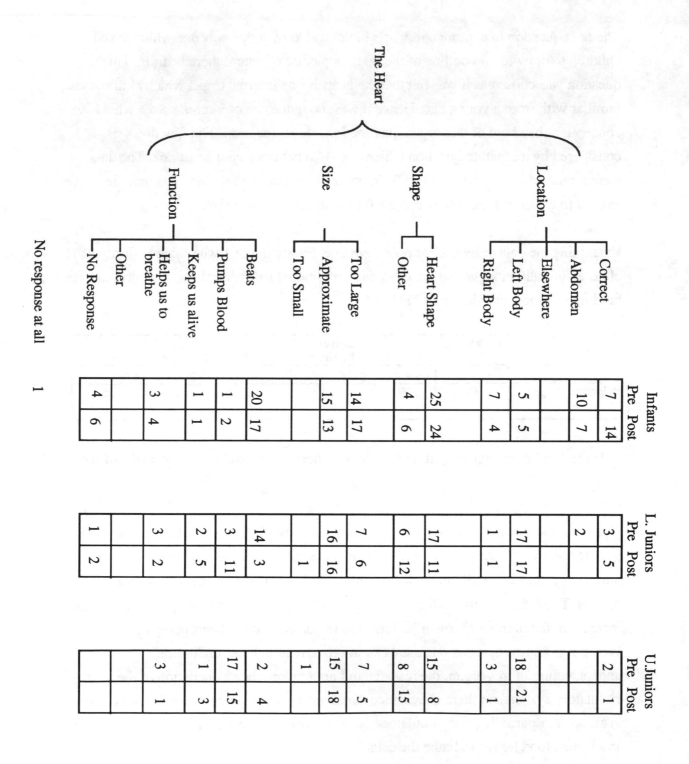

Fig 5.3 Network showing children's responses to questions about the Heart

A similar change in the numbers indicating that the heart pumps blood was not observed for upper juniors as the majority of these children gave this response in the pre-intervention elicitation. This difference between the two groups in the pre-elicitation phase is significant ($p<0.01$). This may be indicative that this change in understanding of the heart would have developed in time anyway. The effect of the intervention was simply to accelerate the change.

The other significant ($p<0.05$) difference prior to the intervention was between the infants and the lower/upper juniors. A large number of the former located the heart centrally in the abdomen. Only a very small number of the latter group indicated such a location. Interestingly, more infants indicated a central location for the heart pre- and post-elicitation than any of the other two groups. The reasons for this were not explored and the only surmise is that children of this age, sense the heart as being central, but cultural messages eventually predominate in a situation where the object in question is not available for inspection.

The picture that emerges from the data is that the overwhelming majority of children see the heart as a valentine shaped object, located on the left of the body with a size which is approximately similar to its real size. Their conception of the function varies but none were able to articulate a view more comprehensive than a knowledge that it 'pumps blood'.

Whilst the origin of this conception of its shape can be ascribed to a wide range of the media, where it is reinforced daily, there is a need for primary science education to recognise the prevalence of this idea and provide children with the scientific view.

The final question which considered what knowledge children had of their own bodies was one where they were asked to add to an outline and 'draw a picture to show what else is inside your body'. Typical examples of the responses have been discussed previously and the data are shown in table 5.10.

An examination of the data in table 5.10 shows that the three internal parts of the body shown predominantly by all children of all age ranges were the heart, bones and the brain (with the exception of infants prior to the intervention). The average number of organs shown increased across the age range and between the pre- and post elicitation for both infants and lower juniors which would indicate that the intervention has had some success in improving children's knowledge.(Fig 5.4).

	Infants		Lower Juniors		Upper Juniors	
	Pre	*Post*	*Pre*	*Post*	*Pre*	*Post*
Blood	9	6	3	3	1	1
Bones	25	24	13	10	12	9
Heart	16	20	18	20	19	18
Stomach	10	9	6	5	12	4
Belly	5	6	1	2	2	0
Brain	1	12	9	10	18	16
Kidney	1	0	1	3	5	10
Liver	0	0	0	2	3	4
Lungs	0	8	5	9	11	14
Windpipe	0	6	4	8	10	9
Bladder	0	0	0	0	1	2
Other	0	1	1	1	4	4
Veins	1	4	5	8	8	7
Intestines	0	1	0	1	3	8
Guts	7	5	2	6	0	1
Muscles	4	0	1	0	5	7
Food	0	0	1	0	0	0
Eyeball	0	0	0	0	1	0
Nerves	0	0	0	0	0	0
Totals	79	102	70	88	115	114
Av No	2.72	3.52	3.0	3.8	5.0	5.0

Table 5.10: Numbers indicating Internal Organs present in the body

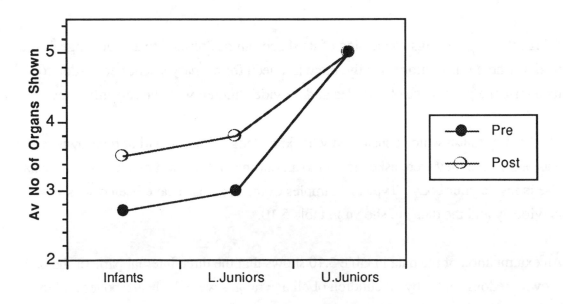

Fig 5.4: Average number of internal parts of the body shown by each age group, pre and post-elicitation

A similar trend was found by Gellert (1962) who also found that the most frequently named parts were bones, blood, the heart and brain. Her test was slightly different in that she asked children to name parts of the body rather than draw them. The difficulty in representing blood in drawings may account for the absence of blood in the data shown above.

This data would also confirm Carey's (1985) evidence that children's biological knowledge develops between the ages of 5 and 11. The effect of the intervention was positive for both the lower juniors and the infants but had no effect on the upper juniors. It is possible that the average number of body parts indicated by upper juniors represents a plateau which is not crossed until several years later.

Another index of the improvement is the number of organs mentioned by three or more children (Table 5.11)

Age Range	No of Organs	
	Pre	*Post*
Infants	7	10
Lower Juniors	8	10
Upper Juniors	11	12

Table 5.11: No of children indicating 3 or more organs/parts of the body.

With reference to particular parts of the body, the only significant change after the intervention was the number of infants indicating the presence of the brain ($p<0.01$) and lungs ($p<0.01$). Both of these increased after the intervention. Again it is notable that prior to the intervention, infants differed significantly from upper and lower juniors whilst afterwards they did not. This would suggest that the effect of the intervention has been to accelerate a natural occurring development in children's biological knowledge.

What processes are performed by components of the body?

This question was explored through a range of questions. In essence the question is a superordinate one to the question of what are the parts/organs of the body since the processes undertaken by the body can only be explained by a child when he or she has a knowledge of their components and their interactions. Hence breathing (gaseous exchange) is a process where the structure of the lungs enables oxygen, a component of the air, to diffuse into the blood stream in the many capillaries which are found in the lungs. Whilst such an answer would not be expected from a primary age child, it illustrates the point that processes are descriptions of interactions and dependent on a basic description of the ontological nature of the body.

The first question asked of children was simply "Why do you need to eat?". The question elicited a range of straightforward answers which are shown in Table 5.12. These responses indicate that the children saw eating simply as a life support mechanism in broad terms, the predominant response at all ages being that food is necessary to stay alive. Only four responses from Upper Juniors indicated that there is any component of food which is essential for body maintenance i.e provides vitamins or helps your heart. The data show no evidence that there has been any significant change as a consequence of the intervention, though it may be that the question is in itself broad and failed to elicit the specific discussion of the components of food and their individual ideas.

	Infants		Lower Juniors		Upper Juniors	
	Pre	Post	Pre	Post	Pre	Post
Stay alive	11	9	7	6	10	13
To grow	8	11	8	7	6	7
To keep fit/strong	7	5	5	12	7	8
To keep healthy	8	8	7	0	8	5
To provide vitamins	0	1	1	0	1	2
The food helps your heart	0	0	0	0	0	2
Other	0	1	0	0	0	0
No response	1	2	1	2	0	0
Total number of responses	35	37	29	27	32	37
Mean number of responses per child	1.2	1.3	1.3	1.2	1.4	1.6

Table 5.12: Data showing number of each type of response by children to the question 'Why do we need to eat?'

The most notable feature was the remarkable similarity across the ages. A possible explanation is that these children are operating with a phenomenologically intuitive knowledge based on simple broad mechanisms i.e. that you need food/blood to keep you alive. This area was not extensively covered by the intervention which simply attempted to introduce a system of food classification which associated types of food with a specific purpose e.g. energy giving. It would suggests that children's vitalistic explanations were seen by them as being comprehensive and adequate since essentially many are correct. In fact, it can be argued that any better understanding of the processing of food and the function of blood requires an appreciation of the particle nature of matter and the transformation of substances. Neither of these concepts are generally addressed in primary science and, some would argue are not available to the cognitive processing of such children.

The following questions attempted to explore what children saw as the function of blood and how it moved around the body. Children were asked 'What does blood do?' which was followed by the question 'How is the blood carried around the body?'. The responses to these two questions were analysed using a network (Fig 5.5 and Fig 5.6).

The data show that very few children were incapable of providing any response at all to these questions, though many only attempted to answer one rather than both. Most children that attempted to provide an answer did so in general terms - blood is needed to keep you alive or healthy, or it moves around the body. The latter answer could be considered a more sophisticated answer in that it recognises that blood is a fluid which does circulate.

Children's answers to the question about how blood is carried around the body showed that some children had greater awareness of specific parts i.e. the heart or veins/tubes.

	Infants		Lower Juniors		Upper Juniors	
	Pre	Post	Pre	Post	Pre	Post
% of children mentioning heart or veins/tubes	31	31	56	74	61	65

Table 5.13: Percentage of children mentioning heart and/or veins

The data in Table 5.13 show that there was little change as the result of the intervention as none of the changes were significant. Again, there was a significant distinction between infants and lower juniors/upper juniors. The data show that the intervention did improve the biological knowledge of these two groups though only marginally in the case of upper juniors. The major difference existed prior to the intervention and would support the case that there was some development in children's biological knowledge during the transition from age 5/6 to age 8/9 and this is reflected in the data.

A more detailed examination of the data shows that only three changes were significant as a result of the intervention. At the 1% level of significance, the intervention had the effect of substantially increasing the number of lower juniors who said that the heart was responsible for circulating blood around the body and the number of upper juniors who expressed the idea that blood runs around the body. At the 5% level of significance, there was a corresponding decrease in the number of upper juniors who expressed the view that the function of the blood was to keep you alive. None of the other changes were found to be significant.

The other analysis of the data undertaken to these responses was to examine how many pupils expressed an answer which gave both a mechanism for the circulation of the blood, and a function or purpose for the blood in the body (Fig 5.6). In each case where a response was provided, it was coded as being correct or incorrect from a scientific perspective. Thus responses that said that the heart made the blood go round or that the blood moved through veins were coded as being a correct indication of the mechanism of circulation. Similarly a response that indicated that blood moved around the body was taken as a correct indication of purpose of blood. In making the latter categorisation, it is assumed that this level of knowledge shows an awareness that the blood is transported to all parts of the body. Responses that stated that the function of the blood was to keep you alive were termed incorrect on the basis that such a general statement failed to show any awareness of what might be happening internally at the microscopic level within the body.

The data are summarised in Fig 5.6. and show that the majority of infants and upper juniors gave a response which indicated both a mechanism and purpose for the blood. Only in a small minority could both of these be considered to be correct. Most of the other responses provided a mechanism or a purpose but not both. Only a few children gave no response whatsoever. An analysis of the data show that there were no significant changes as a result of the intervention which is consistent with the picture presented by the data in Fig 5.5.

Question 6 attempted to explore children's knowledge of respiration through exploring whether their responses gave any indication of gaseous exchange. A wide variety of responses was provided to this question and a simple classification schema was adopted which summarised the responses (Table 5.14).

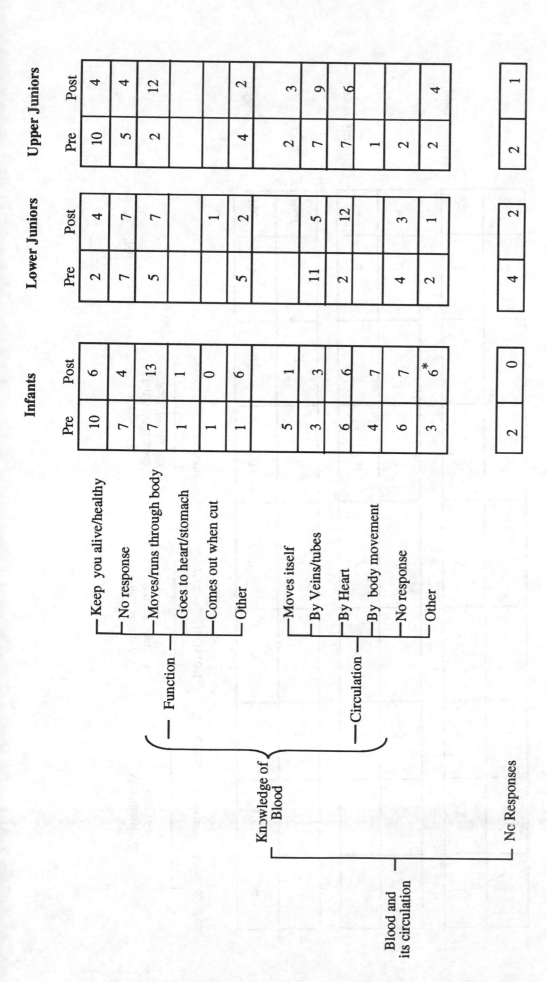

Fig 5.5: Network for responses on the nature and function of blood

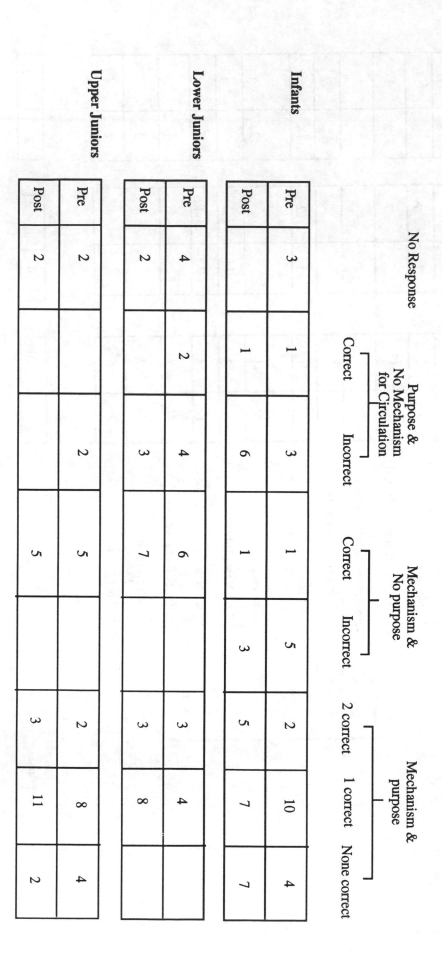

Fig 5.6: Network showing Number of children's
on the purpose and mechanism of blood and their nature

Essentially the schema represents the researchers' attempts to classify the data into scientific, partially scientific, everyday and other explanations. Using this schema the data obtained from children is shown in Table 5.15.

Responses indicating an awareness of the process of respirations	*Responses indicating a possible understanding of some aspect of respiration.*	*Everyday responses*	*Unclassifiable/Other*
Air comes down, carbon dioxide comes out. Turns into carbon dioxide.	It goes into your lungs. It gives air to the blood. The good bit stays in, the bad bit goes out. We breathe in oxygen.	Helps us to breathe. It goes into your tummy. Goes into your mouth and out of your nose	You can smell the air

Table 5.14: Schema used for categorising responses to the question 'What happens to the air which we breathe in?'

	Infants		Lower Juniors		Upper Juniors	
	Pre	Post	Pre	Post	Pre	Post
Responses indicating an awareness of the process of respirations	0	0	2	0	1	2
Responses indicating a possible understanding of some aspect of respiration.	2	5	6	10	9	13
Everyday responses	20	20	8	6	8	4
Unclassifiable/Other	2	1	0	1	0	1
No response	5	3	7	6	5	3

Table 5.15: Data for children's responses on the purpose of breathing

The general pattern shown by these data is one where little change is apparent. There was some minor improvement in the responses which showed some understanding of aspects of respiration but none of these changes have any statistical significance. What is noticeable is that there is a significant ($p<0.05$) diminution in the number of everyday responses between the infants and the other two groups prior to the intervention. The intervention sustains this difference and these data reinforce other data which also showed significant differences existing naturally between these two groups.

The lack of any improvement in children's understanding raises the question of whether this was due to the failure of the intervention or whether the topic is inherently too difficult for children of this age. This issue will be explored later in greater depth.

The next question to address children's knowledge of processes performed by the human body was a question which asked them to add to the drawing to show what happened to food and drink inside your body. The responses to an ostensibly simple question were extremely revealing in what they indicate about children's understanding of their own bodies and in turn, those of other animals. The data were analysed using a network shown in Fig 5.7.

The most noticeable feature was the large number of infant children who failed to show any kind of tube connecting the mouth to stomach/belly. The food was shown intact within the body and simply spreading throughout in an undigested form by a mechanism which was not understood by the children. For infant children, the effect of the intervention was to significantly ($p <0.01$) increase the number of children who showed a tube between mouth and stomach. This represented a positive achievement of the intervention as no understanding of digestion can be achieved until a child begins to assimilate and appreciate the internal connectivity of the parts and organs of the body. No significant changes were found for lower or upper juniors.

A close examination of the data shows that there was a significant difference ($p<0.01$) in the numbers showing a connection between the mouth and the stomach between the infants and lower juniors prior to the elicitation. This suggests that some change in their understanding of children is generated by normal life experiences. The effect of the intervention would appear to be an acceleration of such a change for infants in knowledge and understanding. However, the intervention failed to correct the assumption which was held by a substantial minority of lower juniors and upper juniors that there are two tubes involved in the process of digestion. Whether a

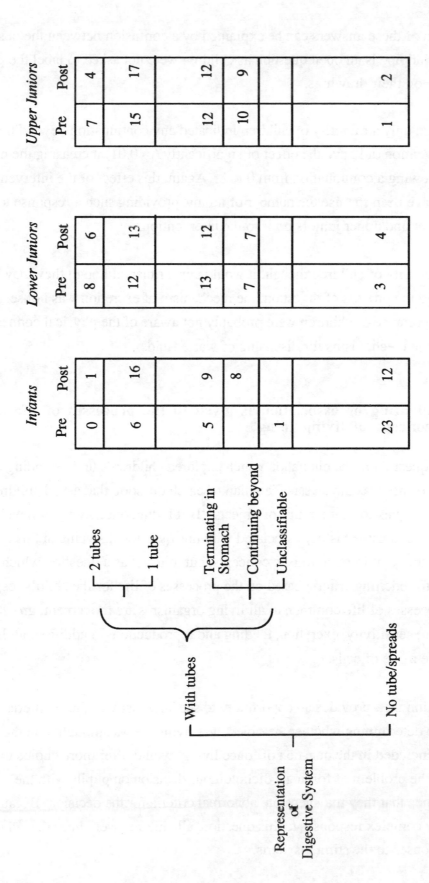

		Infants		Lower Juniors		Upper Juniors	
		Pre	Post	Pre	Post	Pre	Post
2 tubes		0	1	8	6	7	4
1 tube		6	16	12	13	15	17
Terminating in Stomach		5	9	13	12	12	12
Continuing beyond		1	8	7	7	10	9
Unclassifiable							
No tube/spreads		23	12	3	4	1	2

Fig 5.7: Network showing nature of children's responses about the process of digestion.

proportion of these answers can be explained by a confusion between the oesophagus and the windpipe is an open question as children were not asked to label the parts that they drew on their drawings.

In all cases, only a minority of children indicated any continuation beyond the stomach. The intervention did have the effect of significantly ($p<0.01$) increasing the number of infants showing a continuation from 0 to 8. Again, this effect of the intervention would seem to have been to raise the number of infants providing such a response to the level of the lower and upper juniors prior to the intervention.

For the majority of children though, it would appear that, although they may have assimilated the process of digestion, the mechanism of excretion was not seen as being one of importance or children were probably not aware of the physical connection for solids or the mechanisms for disposing of waste fluids.

What weighting or association is given to the processes of life in the child's concept of living thing?

The final question in the elicitation which explored children's understanding of the processes of life was the classic Piagetian research question that asked children to consider the question for a range of objects, 'Is it living, once living or never living?'. Despite the extensive history associated with this question, its value in this context was not to explore children's animistic concepts, but to use it as a question which had potential for eliciting criteria based on the processes of life for the child's response. The seven processes of life common to all living organisms are movement, growth, respiration, sensitivity, excretion, feeding and reproduction. In addition, all living objects are made of cells.

The question then provides an opportunity to see how many of these criteria were applied to determining whether the object was living. It was thought that the extra category included in the question of 'once living' would offer more choice to pupils and diminish the problem of forcing a dichotomous decision on pupils with the consequence that they may generate abnormal criteria for the decision. Because of the inevitably complex responses to this question, all children were interviewed to obtain their responses to the stimulus items.

The objects/pictures shown to children were three clearly inanimate objects by scientific criteria and the normal criteria adopted by adults - a plastic box, a rock and a spoon; three clearly animate objects - a plant, a mammal and an insect and three objects where the decision is more difficult and depends on a full understanding of scientific criteria. These were a toy car which was used to explore whether children were using the simple criteria of movement reported by Piaget and others, an apple and a seed which from a biological perspective are both living organisms but can easily cause confusion.

The criteria used by children were grouped under the headings shown in Table 5.16 which emerged from the data. Table 5.16 shows the major categories i.e. external structure, internal structure, behaviour, tautological and actions and examples of reasoning in each category. The category 'Tautological' refers to justifications that simply appealed to the self-evident, that is explanations that deemed the object 'is alive' or 'is dead' and essentially failed to provide any justification for the assertion. The category of 'Actions' refers to all those justifications which were based on what the animal/object was capable of doing or being used for. In addition, there was one extra category for responses which were unclassifiable.

External Structure	Internal Structure	Behaviour	Tautological	Actions
No face	Has seeds	Movement	Dead/Alive	Can be eaten
Hard	Comes from an egg	Made to move	Was living	Has a use
Broken	Has a Heart	Grows	Don't live	Has to be made
Bent	Has a Brain	Eats		Can be bought
Got hair	Has lungs	Talks		Play with it
Smooth	Has a Liver	Origin		Perform action on object
Rusty/Dirty/Old	Has Teeth	Dies		
Metal	Has a Stomach	Lives in		
Surface feature	Has a Bones	Drinks		
Breaks	Has Blood	Breathes		
Other		It's like		
Too cold		Gives birth		
It's plastic etc		It sees		
Has Legs/eyes		It sleeps		
Solid				
Has a Nose				
Has Ears				
Has Mouth				

Table 5.16: The five categories used for classifying children's responses with all response types listed in each.

The results obtained are summarised using this classification in Table 5.17 beneath. The predominant feature that emerges from the data was the use by children of two broad groups of criteria for their response -that of behaviour, where the major criteria is

	Infants		Lower Juniors		Upper Juniors	
	Pre	*Post*	*Pre*	*Post*	*Pre*	*Post*
External Structure	35%	15%	16%	11%	16%	10%
Internal Structure	0%	8%	0%	2%	2%	10%
Behaviour	38%	51%	59%	70%	62%	67%
Tautological	5%	1%	2%	1%	0%	3%
Actions	18%	20%	9%	9%	13%	4%
Unclassifiable	1%	2%	6%	1%	1%	3%
No Criteria	2%	3%	8%	6%	6%	4%

Table 5.17: Table showing percentages of children's responses in each category.

whether the object is capable of growth and/or movement, and the external structure particularly by infant children. The trends in the changes are perhaps shown more clearly in Fig 5.8.

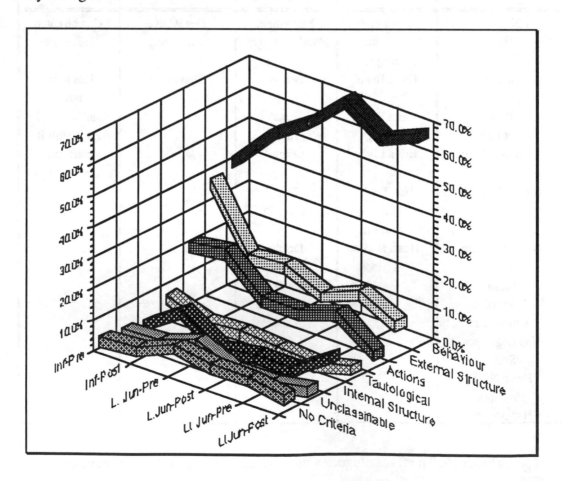

Fig 5.8 Chart showing percentage of each type of response by age group of children

This figure clearly shows that the importance of external structure diminishes and that the pupils attention focus more specifically on the behaviour of the organism as they get older. The predominant criterion used is that of movement which can be seen from the number of responses that mention *specific* processes of life (Table 5.18). All of these are a subset of the category of 'behaviour' defined in Table 5.15 and the data are given to show how often such processes were used as criteria.

	Infants		Lower Juniors		Upper Juniors	
	Pre	*Post*	*Pre*	*Post*	*Pre*	*Post*
Movement	35	70	66	74	57	74
Growth	9	25	39	61	41	61
Reproduction	0	0	1	7	2	7
Digestion	6	6	11	15	13	15
Respiration	0	3	10	14	9	14
Sensitivity	0	0	0	0	0	0
Excretion	0	0	0	0	0	0

Table 5.18: Numbers of responses by children using the criteria of specific processes of life

An examination of the data shows that two changes for the infants pre and post-intervention were significant at the 1% level. These were the diminishment in the response mentioning external structure and the increase in the number of responses falling in the behavioural category. An examination of responses that were specific to the processes of life showed that part of the contribution to the significance was the increase in the number of responses using movement as a criteria ($p<0.05$) and growth ($p<0.05$). Significances were not calculated when responses in any one category were less than 10.

It should be noted that there is a significant difference between infants and lower juniors prior to the elicitation in the number of the responses they give mentioning external structure (Table 5.17) - infants giving many more. This would suggest that the change that occurred as a result of the intervention was simply an acceleration of an event which happens naturally.

The two other significant changes in the data were the increase in the number of responses based on the behaviour of the organism ($p<0.01$) by lower juniors, and the decrease in the number of responses based on actions by upper juniors ($p<0.01$). Table 5.18 shows that the former change is explained by the larger number of responses given by lower juniors which mention the process of growth as a criterion of judgement and the latter change simply by a reduction in the number of upper juniors who mention actions (Table 5.17).

The results shown here give some support to the work of Piaget and others which indicate that the predominant criterion deployed by children was that of movement. However, what they show in addition, is that these children used a variety of criteria, particularly those based on external structure. The data also show some evidence that an increasing number of children use scientific criteria as a basis for judgement. These data then would support the work of Lucas[4] et al (1979) who found similar results in their work. Hence like Lucas, we would argue that such work has ignored the 'richness of children's responses' to this complex question and attempted to adjust the data to fixed categories which our data can not support.

A further analysis of the development across the age range was obtained from examining the number of statements used by children to decide on whether an object was living, once living or never living; the range of categories of criteria that they use and the number of statements which they get scientifically correct. The data are shown in table 5.19.

	Infants		Lower Juniors		Upper Juniors	
	Pre	*Post*	*Pre*	*Post*	*Pre*	*Post*
Total No of criteria used.	279	291	235	272	214	296
Total No of Categories of Criteria used.	81	84	58	54	57	51
No of objects correctly identified as alive	121	150	135	146	144	149

Table 5.19: Data obtained on total No of criteria used, No of Categories and No of objects identified as alive for the whole sample.

The clear trend is that the total number of reasons provided for their choice remained fairly constant, but the number of categories of criteria used diminish from infants to lower juniors. This is shown more clearly by an examination of the averages for the groups (Fig 5.9). The broad trend shows that the reduction in the number of categories

---

[4]   Lucas, A.M, Linke, R.D & Sedgwick, P.P (1979)   Schoolchildren's criteria for "Alive": A content analysis approach. *The Journal of Psychology*, 103, 103-112

used was accompanied by an increase in the number of criteria and an improvement in making the correct scientific judgement of whether an object was living or not.

What is clearly missing from children's understanding at any level was any recognition of excretion or sensitivity i.e. response to stimuli as being a criteria for determining whether an object was living or dead. These were not processes generally perceived by children or projected to other organisms. Either this is because the processes are difficult to observe easily, which hardly seems possible, or more likely, that such processes have not been drawn to the attention of children by their teachers and parents.

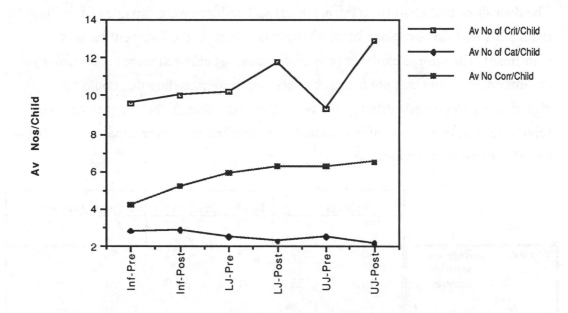

Fig 5.9: Chart showing variation in Average No of criteria; average No of correct decisions as to whether the object was living, once living or never living and average No of categories used.

What the child's knowledge of plants and their parts?

Only one question was used to explore this aspect of their knowledge where children were asked to label a drawing in both the pre- and post-elicitation. The limited exploration of this aspect was in part a reflection of the apparent emphasis within the National Curriculum. In the then version (Science in the National Curriculum, 1989), these stipulated that children at Level 1 of understanding should 'be able to name the external parts of the human body/plants, for example, *arm, leg/flower, stem*' and at level 4 should 'be able to name the major organs and organ systems in flowering plants and mammals'. Hence the intention behind the question used was simply to examine whether children were at all capable of meeting the requirements of level 1.

Traditionally, science education has given little emphasis to this domain of knowledge, a fact which has led one commentator to lament 'Where have all the flowers gone?'[5]. Hence it was decided to simply use a diagram of a flower and ask children to name the parts indicated with a label. The four parts were the flower (or petals), the leaves, stem and roots. Answers that used the correct name or an appropriate terminology i.e. petals for flower, or stalk for stem were coded as being acceptable. Data obtained are shown in Table 5.20.

The data show that apart from infants, nearly all children were capable of labelling the main parts of a flower. None of the changes observed in the intervention was significant. The simple explanation is that since most children were successful prior to the intervention this was not a area of conceptual understanding that could be significantly improved, although the table does show that all the changes bar one represented an improvement in the number of children who were capable of giving a sensible or correct answer.

		Infants		Lower Juniors		Upper Juniors	
		Pre *%*	*Post* *%*	*Pre* *%*	*Post* *%*	*Pre* *%*	*Post* *%*
Flower	*Right/ sensible*	90	77	87	91	96	87
	Incorrect	10	23	13	9	4	13
Leaves	*Right/ sensible*	83	97	100	100	87	96
	Incorrect	17	3	0	0	13	4
Stem	*Right/ sensible*	40	60	91	91	83	87
	Incorrect	60	40	9	9	17	13
Roots	*Right/ sensible*	33	43	83	96	91	91
	Incorrect	67	57	17	4	7	9

Table 5.20: Data for responses to question asking children to label parts of a flower.

Since the time that the research was undertaken, the National Curriculum Order has been changed to specify that it is the organs of the plant that children should be able to name. To some extent, this question shows children's capacity to name the parts but a

---

5   Honey, J. (1987) Where have all the flowers gone? *Journal of Biological Education*, 21, 3, 188-189.

more detailed question would have been required to explore the present needs of the National Curriculum.

6: Conclusions

In the research reported here, there are a number of notable features and patterns which this section draws attention to. Essentially these are:

a)   The distinction in the level of understanding and knowledge between infants and lower/upper juniors, prior to the intervention;

b)   The raising of the level of understanding of the infants as a consequence of the intervention;

c)   The lack of improvement by lower and upper juniors of their understanding and knowledge;

d)   The implications for the current National Curriculum science order.

Clearly identifiable within the data prior to the elicitation is a distinction between the level of knowledge and understanding shown by the infants and the other two groups. Infants were weaker at successfully discriminating 'healthy' foods from 'unhealthy' or 'indeterminate' foods (Table 5.2) and similarly less able to discriminate 'healthy' activities from 'unhealthy' activities (Table 5.4). In both instances, the weakness was a tendency to respond that 'unhealthy' activities were 'healthy'. In part, table 5.5 would suggest that there is a failure on the part of infants to recognise that exercise and sleeping are 'healthy' activities as only 14% produced drawings with these activities in in response to question 9 (Appendix 2) as opposed to 61% of lower juniors and 57% of upper juniors.

Again, infants' responses to the question asking them to draw a 'healthy' and an 'unhealthy' meal showed that their knowledge was distinctive from the other two groups. They were much less consistent in their views, prioritising a different range of foods as being healthy and with much lower percentages consistently drawing these foods (Table 5.6). Whilst peas figure in the list of healthy foods for all groups, only 17% of infants recognised that vegetables could be considered healthy as opposed to 74% (lower juniors) and 61% (upper juniors). Strangely, 34% of infants did mention fruit but this type of food was only mentioned by a small minority of lower (13%) and upper juniors (9%).

A similar trend was found in the responses for an 'unhealthy' meal. 83% of both lower and upper juniors drew chips but only 28% of the infants. For them the most unhealthy food were sweets (79%) mentioned by only 17% of lower and 9% of upper juniors. The evidence from these data, and those in the preceding paragraph, would suggest that the infant group did not have a clearly defined notion of what is a 'healthy' food and they were failing to make the common distinction between food and snacks.

In terms of their biological knowledge, infants on average drew the lowest number of internal parts/organs in the body (2.7), compared to lower juniors (3.0), and upper juniors (5.0). Whilst this is not significant, it does show a clear trend which is supported by other data. Prior to the intervention, significantly more (p<0.05, Fig 5.3) infant children located the heart centrally in the abdomen than the other two groups of children, and significantly fewer infants (p<0.05, Table 5.12) mentioned the heart or veins in explaining how blood is moved around the body. Significantly more infants (p<0.05, Table 5.14) also gave everyday responses in explaining respiration, as opposed to responses that showed a possible understanding or some definite understanding.

Only 20% of infant children drew any physical connection between the mouth and the stomach to explain digestion as opposed to 87% of lower juniors and 96% of upper juniors (Fig 5.7). In deciding whether an object was living, not living or once living, infants made much greater use of criteria based on external structure than older children and much less use of behaviour (Table 5.17) and this difference was significant (p<0.05). Only 20% of infant responses used criteria based on specific processes of life as opposed to 54% for lower and 57% for upper juniors. In responding to this question they used a larger number of criteria and were less successful at identifying which objects were living, non-living or once-living than lower or upper juniors. Finally in labelling the parts of a flower, the infants were the only group where less than 80% correctly labelled all the parts of the flower.

Taken in its entirety, these data would suggest that there were significant differences prior to the intervention in the level of understanding and knowledge of infant children with a median age of 6 compared to children 2 years older. Given that this domain of knowledge is not normally an area of specific treatment within schools, the data would suggest that the epigenesis of such knowledge is a naturally occurring event between the age of 6 and 8. The emphasis by infants on external features as a criterion for deciding whether an object is living, once-living or not living, coupled with this group's failure to envisage what lies inside the body, provides some evidence to support the Piagetian argument that the improvement reflects a development in the child's general capabilities to perform mental processes. That is, young children centrate on the concrete and observable and lack the imagination to envisage entities which remain for them hypothetical i.e. the heart, lungs. However, at a simpler level, it is much easier to explain that what the data reflects is a growth in the child's awareness and domain-specific knowledge and the former argument is at best a tenuous hypothesis.

Effect of the Intervention

An analysis of the data shows two contrasting effects of the intervention. Firstly that there are many instances when the knowledge and understanding of the infants has been improved and secondly, that it failed to make any significant effect with lower and upper juniors.

Infants understanding of the location of the heart improved as a result of the intervention with a drop in the number indicating that it was located in the middle to low abdomen, and an increase from 24% to 48% in the number showing it in the correct location.

Fig 5.4 shows that infants showed the greatest increase in the average number of organs/parts of the body drawn per child from 2.7 to 3.5. Particularly notable was an improvement in the number of infants who drew the brains and lungs after the intervention whilst the other two groups showed little or no improvement in the numbers indicating these organs (Table 5.10). The number of organs/parts drawn by lower juniors also improved but the upper juniors showed no improvement whatsoever, suggesting that a plateau of knowledge may be reached. Also infants were the only group to show a significant increase ($p<0.05$) in the number mentioning that muscles could be found everywhere in the body. The data would suggest that this domain of knowledge is amenable to development with appropriate teaching and learning strategies, particularly for infants.

In explaining what happened to food and drink inside them, infants showed a significant ($p<0.01$) increase in the number who included a tube or tubes between the mouth and a central body cavity (Fig 5.7), whilst there were no significant changes for lower and upper juniors.

Infant children's justifications for their choice of whether an object was living, once living or never living showed a significant decrease ($p<0.01$) in the number using an external structure as a criterion and an accompanying significant increase in the number using behaviour after the intervention. Most of the latter effect can be explained by the increase in the number who used the criterion of movement. The latter criterion, and that of growth, constituted only 33% of all the criteria used. Such a figure accompanied by the wide use of other criteria does call into question, the simple Piagetian schema for children's reasoning described in chapter 1.

Other significant changes are somewhat random. Significant at the $p<.01$ level was an increase in the number of lower juniors who mentioned growth as a criterion for deciding whether an object was alive. For upper juniors there was a significant reduction in the number who mentioned actions as a criterion. Lower juniors responses to the question asking for four things to do with keeping healthy showed a significant ($p<0.01$) increase in the number of drawings which showed exercise. The questions on the function and purpose of blood showed that there was a significant ($p<0.01$) increase in the number of lower juniors who said that the heart was responsible for circulating the blood.

Upper juniors showed a significant ($p<0.05$) decrease in the number who showed the heart as a valentine-shaped object and an increase in the number who explained that blood runs around the body. Accompanying the latter change was a decrease in the number of everyday explanations which stated that blood was to keep you alive.

Overall, this would suggest that such an intervention had little effect for older children but there was a much more positive effect for infant children. One explanation would be that none of the knowledge and understanding explored by the intervention demanded radical restructuring of the child's knowledge, but simply an addition or accretion to an incomplete or everyday understanding. Hence, exposure to such knowledge produced many significant changes when first explored with infants, but had little effect with children who are older and may already have a reasonable level of knowledge in the domain.

Implications for the National Curriculum

The latest version[1] of the National Curriculum science order expects that pupils should develop knowledge and understanding of 'life processes and the organisation of living things'. Achievement of this aim will be measured by a set of statements of attainment defined from level 1-10 of which only those from level 1-5 are appropriate for children

---

[1]   Department of Education and Science. *Science in the National Curriculum*. HMSO. 1991

in primary schools. Those statements of attainment which are relevant to this attainment target are shown below.:

Level 1   (a) be able to name the external parts of the human body and a flowering plant.

Level 2   (a) know that plants and animals need certain conditions to sustain life.

Level 3   (a) know the basic life processes common to humans and other living things.

Level 4   (a) be able to name and locate the major organs of the human body and the flowering plant.

Level 5   (a) be able to name and outline the functions of the organs and organ systems in mammals involved in circulation and reproduction and those in flowering plants involved in sexual reproduction.

Most of these statements of attainment are open to considerable interpretation but some exemplars are provided for additional meaning. The level 1 statement expects children to be able to name such parts as ' stem, leaves and petals'. The data in table 5.20 would suggest that at least 80% of lower and upper juniors would have little difficulty in achieving this. However only 40% of infant children were able to appropriately mark the stem when asked.

The current level 2 statement which expects pupils to be able to describe how to look after a pet animal or a potted plant was not explored by this research.

The level 3 statement essentially expects children to be able to identify six of the seven processes of life (excretion is excluded), and to recognise that these are common to themselves and familiar animals. The data gathered here would suggest that only a minority of children are likely to attain such a level of knowledge and understanding. Whilst movement and growth are the most commonly recognised, these criteria were only used at best by 46% of upper juniors in deciding whether an object was living, once living or never alive. It may be argued that such a question is not the most appropriate for eliciting such information from children. However, such questions are standard methodological instruments which have been used and tested for their reliability. Variations in the questions such as that used by Lucas et al[2] where the child was shown a photograph of a indeterminate object and asked to explain how they would tell if it was alive produced little better in the way of understanding. They found that only a maximum of 26% of primary age children mention 'breathing' as a criterion for deciding. Consequently these data would suggest that such a level of attainment is too high an expectation of a top infant and unlikely to be achieved by even the average upper junior.

The level 4 statement expects children to be able to 'name and point to the approximate positions of organs such as heart, lungs, stomach and kidneys in humans, and stamens and ovary in a flowering plant.' The data in table 5.10 gives the maximum percentage of children in age group who identified each of these organs in their drawings and gives some insight into whether such an expectation is obtainable.

Organ	Infants %	Lower Juniors %	Upper Juniors %
Heart	69	87	78
Lungs	26	39	67
Stomach	65	30	60
Kidneys	3	13	43

Table 6.1: Maximum percentage of children who indicated each organ in their drawings

Assuming that the figure for the number of lower juniors indicating the stomach was an aberration, this attainment target would look to be achievable by the majority of children as long as questions are restricted to basic organs and do not attempt to elicit where such organs as the kidneys and liver are located. If so, the likely chance of children achieving such a level of attainment would diminish significantly.

Only data for children's ability to locate the heart were collected. A maximum of 48% of infants, 22% of lower juniors and 9% of upper juniors correctly indicated the position of this organ. Although this is an organ of which knowledge of its location and shape is particularly susceptible to misconceptions, these data would suggest that many children would again experience difficulties in attaining such a level of attainment which is defined at the average level of attainment for an 11 year old. No data were collected for children's knowledge of the organs and parts of plants.

---

[2]   Lucas, A.M., Linke, R.D. & Sedgwick, P.P. (1979) School children's criteria for "Alive": A content analysis approach.*The Journal of Psychology*, 103, 103-112.

At level 5, children are expected to 'explain how the heart acts as a pump in the body to circulate blood to the organs'. Such a level of attainment is to be achieved by able, top juniors and the data in Fig 5.3 show that a maximum of 74% indicated that the heart 'pumps blood'. However, this does not show that they understood that there is a double pattern of flow (systemic and pulmonary). In addition, our question failed to explore the models of the circulation system held by pupils and the work of Mintzes et al[3] suggest that many children do not hold closed circulatory models which may be implicit in this attainment target.

The conclusion that can be drawn is that attaining this level of knowledge and understanding is not a simple matter for many children, and that in particular, the knowledge associated with level 4, or even level 3, is not held by many children at the age of 11 at which they will be tested. It is also possible that the knowledge expected at level 5 is less demanding for children and that some of the statements of attainment should be reversed. Since these statements represent empirical definitions based on 'expert's' intuitions, this would not be surprising. Fuller sets of data from the trials of the standardised assessment testing conducted in the summer of 1992 may support such a hypothesis.

3   Mintzes, J.J, Trowbridge, J.E. & Arnaudin, M.W. (1991) Children's Biology: Studies on Conceptual Development in the Life Sciences in Glynn, S.M, Yeany, R.H. & Britton, B.K. (Eds) *The Psychology of Learning Science*. New Jersey: Lawrence Erlbaum

Appendix 1: Schools

ILEA

Inspector for Science Education   John Wray

Schools

Ashmole Primary

St Edmunds Roman Catholic School

Brampton School

Culloden

Summerside

Altmore

Appendix 2: Elicitation Questions

The following questions were used for the elicitation activities with children

1.   Which of these are healthy foods? (Please ring)

     Lettuce   sugar   bread   meat   chips

        orange juice   apples   rice   potatoes

     burger   crisps   biscuits

2.   Which of these are to do with keeping healthy? (Please ring)

     running   arguing   watching TV   feeling happy

        eating   playing with friends   laughing

     swimming   sleeping   smoking   fighting   reading

3.   Why do you need to eat?

4.   What does blood do?

5.   How is blood carried around your body?

6.   What happens to the air which you breathe in?

7.   Where in your body are your muscles?

8.   Keep a diary of all the things you do in one day?
    Which of these are to do with keeping healthy?

9.   Draw 4 things which are to do with keeping healthy.

10. On the diagram, draw a 'healthy meal' and 'a not so healthy meal'

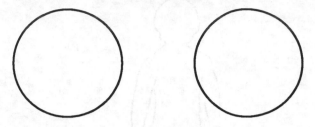

A healthy meal   A not so healthy meal

11. Add to the picture to show what happens to food and drink inside your body.

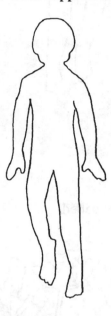

12. Add to the picture to show what else is inside your body.

13.   Can you add to the picture to show where your heart is?

What does your heart do?

14.   What are the parts of this plant called/

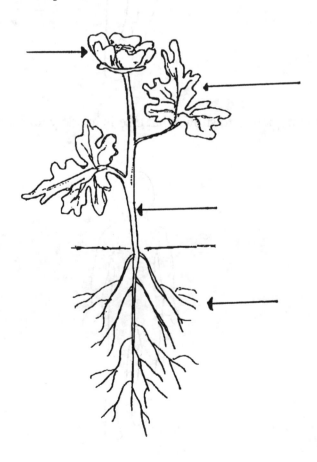

15. For each of the following, say whether it is living, once living or never living

Object	Response	Reason
A plastic box		
A piece of rock		
A spoon		
A plant		
An animal		
An insect		
An apple		
A toy car		
A seed		

Appendix 3: Intervention Activities

This appendix contains the notes which were provided to teachers for the intervention activities. In the briefing provided to teachers, emphasis was placed on using a range of these strategies in any order that suited their work. Teachers were asked to encourage children to use these activities to explore their thinking and understanding of the processes of life and to use these as a focus for generating further pupil investigations.

Possible Interventions

The following is a list of possible strategies for developing children's ideas:

Group discussions - children can be given cards to sequence or to sort as a basis for discussion.

Data bases - these are a useful way of collecting and collating information and displaying it in a clear way so that the children can discuss it. They could take the form of:- sets (Venn diagrams), graphs (block graphs, tick graphs etc.) or computer data bases (Using programs such as 'Ourselves' or 'Our Facts'.)

Sorting/classifying activities - these help children to clarify their thinking, to work co-operatively and to exchange ideas. They could be simple sets e.g. living or non-living, or a more complicated form of classification e.g. Does it Move? Does it live in Water? (use the program 'Branch' -available as part of the MEP pack or as part of 'Junior Ecosoft'). Children can use these to devise their own classification tables or use ones which you devise for them. Logic trees ('decision trees' in the maths document ,page 37), are also a good way of helping children to classify using observable characteristics.

Devise their own investigations to test out their thinking. The children should have plenty of opportunity to devise their own investigations using a wide range of equipment which should be made available to them .

Specific Activities

The following is a list of suggested activities which can be used with particular topics.

Health education

Questions/activities that can be used with children to elicit their thinking

Foods: Draw the meal which you ate last night.

Where did the food come from?

Growth: How do they know that they have grown?

What helped them to grow? (Living things grow)

What goes into your body?

What do these things do? Are all of these things safe/healthy?

Group
discussion: How do healthy people look? What do they do? Make a message
for not so healthy people. e.g. Go to bed early.

Sorting
Activities   Sort the list of words to do with keeping healthy as a group to see if
they can influence each other. e.g. laughing, reading, sleeping

Which of these things do healthy people do - all of the time, some
of the time, never? Make up a healthy day. Give the children cards
to sequence.

Drawing/
Modelling
Activities   Make class graphs to show which foods are healthy. Use as a basis
for discussion. How do we know these things?

Make a class graph of the types of exercise taken. Use this as the
basis for discussion.

Feelings

Group
discussion: What can I do to keep myself:- clean? safe?healthy?
Who else helps?

Drawing/
Modelling
Activities   Make a class poster to show things that make us feel good, things
that make us feel bad. What are good feelings? What are bad
feelings? e.g. feeling sad, lonely, worried

Food

Group discussion:	What happens to the food which I eat? Use cards to sequence this. What connection is there between eating and going to the toilet?
Sorting Activities	Sort foods into fruits, bread and cereals pulses (beans and dried food), meat, vegetables, sweets and cakes.

Which foods are :- fatty? sugary? salty? |

How our bodies work

Investigat-ions	Children could be asked to feel their bones. If they have the opportunity to look at a model skeleton, are they able to feel where most of these bones shown on the skeleton are in their own bodies?

Can they find out how many joints they have?

Look at a large model skeleton and draw it. Find out what the parts are called.

How many groups of muscles can they find? In the hall exercise each muscle group. |
| Drawing/ Modelling Activities | They could make a model to show how their arm bends.

Make a model skeleton. (Learning through Science cards) |

Blood and Heart

Investigat-ions	Can they feel their pulse?

Use a stethoscope to feel their heart beat.

What is the effect of exercise on heart beat and breathing rate? |
| Drawing/ Modelling Activities | Make a model stethoscope |

What is inside your body?

Investigat- *Kidneys:* - Ask the children to do some filtering to show how the
ions   kidneys work.

            Lungs: - Blow up some balloons to find out what their lung
            capacity is.

Group   Hold a group discussion about the children's own pictures of what
Discussion they think is inside them. Are they correct? How would they
            know?

Sorting   Place cut out parts of the body in the correct place on a large outline
Activities of the body.

Seeds

Investigat- Ask the children to germinate some seeds so that they are able to
ions   realise what the various parts of the plant do, particularly the roots.
             (Most of the children in the elicitation phase where unaware of what
             roots were).

Appendix A4

Data for Children's Responses when asked to draw a 'Healthy' and 'Unhealthy' Meal.

The following are the complete data for children's responses when asked to draw a healthy meal. The figures give the percentage of children indicating each type of food.

Table A4.1: Data for Healthy Meal

	Infants		Lower Juniors		Upper Juniors	
	Pre %	Post %	Pre %	Post %	Pre %	Post %
Carrots	21	14	70	78	78	57
Peas	28	21	65	70	26	48
Potato	14	14	22	30	22	13
Spaghetti	0	10	13	13	0	0
Chips	41	14	0	4	22	0
Rice	10	7	0	0	4	4
Lettuce	14	21	17	26	52	43
Tomatoes	17	7	26	22	26	17
Veg	17	28	74	39	61	26
Beans	14	14	0	0	0	4
Meat	3	17	22	17	17	30
Sausages	17	3	0	0	0	0
Burger	7	3	9	4	0	0
Bacon	10	0	0	0	0	0
Chicken	3	0	0	0	0	0
Fish	34	45	48	39	35	35
Eggs	17	17	4	22	13	13
Cheese	14	3	0	4	0	22
Fruit	34	31	13	22	9	43
Bread	41	14	9	4	17	13
Brown Bread	0	0	0	0	0	0
Orange Juice	10	0	0	9	4	4
Drink	14	17	4	0	0	0
Cake/Biscuits	7	0	0	0	0	0
Cornflakes	7	0	0	0	0	0
Milk	7	10	0	22	4	4
Cod Liver Oil	0	0	0	0	4	0
Other	0	7	0	4	9	9

Table A4.1: Data for Healthy Meal

	Infants		Lower Juniors		Upper Juniors	
	Pre %	Post %	Pre %	Post %	Pre %	Post %
Chips	28	31	83	78	83	100
Burger	3	3	57	43	48	78
Eggs	34	7	26	30	43	9
Sausage	0	7	26	48	26	35
Bacon	0	0	13	9	22	13
Beans	3	7	9	4	17	13
Sweets	79	55	17	22	9	17
Cake	34	7	0	0	4	26
Ribena	17	0	0	0	0	0
Sugar	7	21	17	9	0	0
Fish Fingers	3	14	13	9	4	0
Orange	3	0	0	0	0	0
Salt	3	3	0	4	0	0
Crisps	10	10	4	0	0	13
Ice Cream	14	17	0	4	0	0
Gum	10	0	0	0	0	0
Apple	3	0	0	0	0	0
Biscuits	10	0	0	0	0	4
Coke/Fizzy Drink	3	14	26	22	9	0
Other	3	3	4	4	22	26
Beer/lager	0	10	0		0	0
Fruit	0	10	0	0	0	0
Meat	0	10	22	9	4	9
Chicken	0	14	0	0	0	0
Veg	0	7	0	9	0	0
Butter	0	3	0	0	0	0
Milk Shake	0	0	0	0	13	0
Peanuts	0	0	0	0	0	4

Table A4.2: Data for Unhealthy Meal